I0055176

Advances in Polymer Science

Related Titles

Biopolymers Based Advanced Materials
ISBN: 978-0-6482205-4-1 (e-book)
ISBN: 978-0-6482205-5-8 (hardcover)

Functional Polymer Blends and Nanocomposites
ISBN: 978-0-6482205-6-5 (e-book)
ISBN: 978-0-6482205-7-2 (hardcover)

Functional Nanomaterials and Nanotechnologies: Applications for Energy & Environment
ISBN: 978-0-6482205-2-7 (e-book)
ISBN: 978-0-6482205-3-4 (softcover)

Advances in Polymer Technology: Material Development, Properties and Performance Evaluation
ISBN: 978-1-925823-00-4 (e-book)
ISBN: 978-1-925823-01-1 (hardcover)

Polymer Nanomaterials for Specialty Applications
ISBN: 978-1-925823-03-5 (e-book)
ISBN: 978-1-925823-04-2 (hardcover)

Advanced Materials
ISBN: 978-1-925823-05-9 (e-book)
ISBN: 978-1-925823-06-6 (hardcover)

Biofuels
ISBN: 978-1-925823-12-7 (e-book)
ISBN: 978-1-925823-13-4 (hardcover)

Liquid Crystalline Polymers
ISBN: 978-1-925823-16-5 (e-book)
ISBN: 978-1-925823-17-2 (hardcover)

Polymer Nanocomposites: Emerging Applications
ISBN: 978-1-925823-14-1 (e-book)
ISBN: 978-1-925823-15-8 (hardcover)

Advances in Polymer Science

Dr. Vikas Mittal
Editor and Lead Author

CWP
Central West Publishing

This edition has been published by Central West Publishing, Australia
© 2019 Central West Publishing

All rights reserved. No part of this volume may be reproduced, copied, stored, or transmitted, in any form or by any means, electronic, photocopying, recording, or otherwise. Permission requests for reuse can be sent to editor@centralwestpublishing.com

For more information about the books published by Central West Publishing, please visit https://centralwestpublishing.com

Disclaimer
Every effort has been made by the publisher, editor and authors while preparing this book, however, no warranties are made regarding the accuracy and completeness of the content. The publisher, editor and authors disclaim without any limitation all warranties as well as any implied warranties about sales, along with fitness of the content for a particular purpose. Citation of any website and other information sources does not mean any endorsement from the publisher and authors. For ascertaining the suitability of the contents contained herein for a particular lab or commercial use, consultation with the subject expert is needed. In addition, while using the information and methods contained herein, the practitioners and researchers need to be mindful for their own safety, along with the safety of others, including the professional parties and premises for whom they have professional responsibility. To the fullest extent of law, the publisher, editor and authors are not liable in all circumstances (special, incidental, and consequential) for any injury and/or damage to persons and property, along with any potential loss of profit and other commercial damages due to the use of any methods, products, guidelines, procedures contained in the material herein.

A catalogue record for this book is available from the National Library of Australia

NATIONAL LIBRARY OF AUSTRALIA

ISBN (print): 978-1-925823-59-2
ISBN (e-book): 978-1-925823-58-5

Contents

Anish M. Varghese and Vikas Mittal

4.   Polymers and Composites for Optical Applications   75
Liyamol Jacob and Vikas Mittal

5.   Polymeric Self-sensing Materials   101
Haleema Saleem and Vikas Mittal

Nitride: Thermal, Mechanical, Morphological and Electrical Studies
Mona Al Hosani, M. R. Vengatesan and Vikas Mittal

10.   **Polyethylene-Thermally Reduced Graphene**   **241**
      Nanocomposites: Comparison of Masterbatch and
      Direct Melt Mixing Approaches on Mechanical,
      Thermal, Rheological and Morphological
      Properties
      Ali U. Chaudhry and Vikas Mittal

Preface

Advancements in polymer science have led to the development of functional polymeric materials with optimal property profiles and superior commercial relevance. As a result, more and more polymeric materials are gradually replacing glass and metals in a wide variety of applications. In fact, almost all domains of materials have been significantly influenced by the polymers. The development of polymeric hybrids by incorporating a large number of organic and inorganic phases leads to the further enhancement of their application spectrum. In this respect, the current book presents a variety of functional polymeric systems developed in the recent past with a potential of advanced applications in diverse areas.

In Chapter 1, the effect of amine functionalized polyolefin compatibilizer and graphite oxide nanofiller on the mechanical as well as thermal properties of polypropylene has been investigated. In Chapter 2, polypropylene nanocomposites were generated using binary filler system containing graphene and surface modified sepiolite. The influence of the mixed filler system on the mechanical, thermal and morphological properties of the composites was studied. In Chapter 3, cyclohexylethylene-co-ethylene (CHEE) copolymer was used to develop polypropylene blends and graphene reinforced polypropylene blend nanocomposites using direct melt blending technique. The ultimate objective of the study was to explore the benefit of graphene to simultaneously enhance tensile modulus, tensile strength, impact strength, thermal stability and thermal conductivity in the presence of the CHEE copolymer. Chapter 4 presents a comprehensive review of the various optical properties of the polymeric materials as well as different polymers applied in various optical applications. In Chapter 5, different sensing techniques as well as structural health monitoring (SHM) technology with respect to polymeric composite materials are outlined briefly. Chapter 6 provides in-depth analysis of the properties and performance of flame retardant polymers. In Chapter 7, the effect of amine functionalized reduced graphene oxide on the crystallization, thermal stability and mechanical properties of high density polyethylene has been studied, using chlorinated polyethylene as compatibilizer. Chapter 8 provides a brief review of the synthesis and properties of the liquid crystalline polymer composites. In Chapter 9, compatibilized polypropylene nanocomposites have been developed with graphene for achieving enhancement in mechanical and thermal properties, along with thermal and electrical conductivities. In addition, boron nitride was also used as reinforcing filler so as to compare its performance with graphene for enhancing the polymer properties. Chapter 10 explores the effectiveness of the masterbatch approach over direct melt mixing method in improving the filler dispersion and resulting composite properties.

The book would not have been successfully accomplished without the support of chapter contributors. The book is dedicated to my family for unswerving support, constant motivation as well as constructive suggestions for improvement.

Vikas MITTAL

1

PP/GO Nanocomposites with Amine Functionalized PE-g-MA Compatibilizer

1.1 Introduction

Polypropylene (PP) is a versatile semi-crystalline thermoplastic polymer with wide range of commercial applications. It presents several benefits like low cost, easy processing, low density and high melting point [1,2]. In spite of its advantages, poor impact resistance of PP, especially under intense conditions such as low temperatures or higher strain rates, hinders the utilization of PP as an engineering plastic [3]. Further, due to the non-polar nature of the resin, it is challenging to obtain the homogenous dispersion of the polar fillers in the PP matrix [4-6].

Various methods have been proposed to enhance the impact performance of PP, mainly based on the incorporation of the soft rubbery phase elastomers, however, these materials impact the stiffness of the polymer negatively [7]. The incorporation of the amphiphilic materials, consisting of both polar as well as non-polar groups, is a promising approach, which results the simultaneous enhancement of the toughness and stiffness, owing to the nanoscale dispersion of the filler phase in the matrix [8].

Among the different nanofillers, graphene, a two dimensional nanomaterial, offers opportunities for generating advanced nanocomposites having optimal micro-mechanical behavior [9]. Incorporation of a small extent of graphene nano-sheets in the polymer matrices has been observed to remarkably enhance the physical properties of the host polymers. The possibilities to functionalize its surface further enhance the potential of achieving simultaneous enhancements in stiffness, toughness and other properties. In recent years, different studies have reported the preparation and structure-property relationships of PP/graphene nanocomposites [10]. Song *et al.* [11]

Haleema Saleem, M. R. Vengatesan and Vikas Mittal*, The Petroleum Institute (part of Khalifa University of Science and Technology), Abu Dhabi, UAE
*Current address: Bletchington, Wellington County, Australia
© 2019 Central West Publishing, Australia

reported the generation of PP/graphene nanocomposites with enhanced thermal as well as mechanical properties. It was observed that the incorporation of about 0.42 vol% graphene resulted in 74% increment in the polymer stiffness, whereas the yield strength was enhanced by 75%. In addition, the thermal oxidative stability of the polymer was also enhanced. In another study, Varghese *et al.* [12] studied the effect of thermally reduced graphene oxide (TRG) on the properties of polypropylene/maleic anhydride-graft-ethylene vinyl acetate (PP/EVA-g-MA) blends. The addition of up to 3 wt% TRG in the PP/EVA-g-MA blend enhanced the tensile strength and modulus, while retaining the impact strength. PP/EVA-g-MA/TRG nanocomposites also exhibited higher electrical and thermal conductivities as compared to PP or PP/EVA-g-MA blends.

In spite of the enhancements in the mechanical properties of the nanocomposites, aggregation and stacking of graphene are common issues hindering the optimal enhancement in performance. These phenomena result in stress concentration and hinder the load transfer from the PP matrix [13,14]. For reducing the cohesive forces between the graphene sheets and for developing specific interactions with the host polymer matrix, the functionalization of graphene is, thus, needed.

Graphite oxide (GO), where the carboxyl and hydroxyl functional groups are covalently bonded to the carbon edges and basal planes, has been employed as filler in various polymer matrices [15,16]. It is generated by the chemical oxidation of graphite. However, it is difficult to achieve the homogenous dispersion of GO in non-polar polymers such as PP, polyethylene (PE) and polystyrene. Further, it also exhibits poor electrical properties due to the disconnected sp^2 network and oxygenated functional groups [17]. To overcome these challenges, surface modification of GO is carried out using the oxygenated functional groups present on the surface [18,19]. For achieving the homogenous dispersion of GO in the PP matrix, *in-situ* reduction in the polymer also presents a viable strategy.

Furthermore, as mentioned earlier, the addition of a compatibilizer also serves to bridge the non-polar PP and polar graphene phases. Shin *et al.* [20] used octyl-triethoxysilane (OTES) as a compatibilizer for PP/GO nanocomposites. It was noted that the mechanical performance of the composites increased due to the stronger interaction between the PP and GO phases. For instance, the fracture toughness increased by 117% by the incorporation of OTES and 1.0 wt% GO. To enhance the compatibility of PP with the polar fillers, PP-graft-maleic anhydride (PP-g-MA) or PE-g-MA have been widely employed as compatibilizers. However, literature studies have also confirmed better adhesion of the amine-functionalized PP/PE than the maleated PP/PE compatibilizer. Kobayashi *et al.* [21] conducted comparative study of the influence of PE-g-MA and amino-functionalized PE for enhancing the bonding between polyurethane and polyolefins. The amine functionalized polymer was observed to exhibit greater adhesion than the PE-g-MA compatibilizer.

Several studies have proved that the blending of PP with PE can improve the impact strength of PP at lower temperature [22,23]. Jose *et al.* [24] examined the mechanical properties, crystallization behavior and phase morphology of high density polyethylene (HDPE)/isotactic polypropylene (i-PP) blends. It was observed that the properties of both polymers related closely to each other. Fel *et al.* [25] observed that the incorporation of a small amount of PE (typically < 20 wt%) can increase the impact strength of the PP blends. In another study, Sanchez-Valdes *et al.* [26] studied the compatibilization effect of PE-g-MA and different NH_2-functionalized PEs on the PE based nanocomposites. The NH_2-functionalized PEs were generated by the chemical reaction of PE-g-MA with a tertiary amine 2-[2-(dimethylamino)ethoxy]ethanol (DMAE) and two primary amines, 2-aminoethanol and 1,12-aminododecane to form the corresponding PE-g-DMAE, PE-g-EA, and PE-g-D12 compatibilizers. The PE-g-DMAE compatibilizer was observed to provide an optimally exfoliated morphology and an agreeable balance between optical, mechanical and thermal properties.

In the current study, the effect of amine functionalized polyolefin compatibilizer and GO nanofiller on the mechanical as well as thermal properties of PP has been investigated. In this approach, an amphiphilic compatibilizer, PE-alt-DA6, is added to the PP matrix, along with GO (Figure 1.1). The amine groups present in the PE-alt-DA6 structure react with the carboxyl as well as epoxy groups present on GO, thus, causing the covalent binding of the filler platelets with the compatibilizer. During this process, GO also gets chemically reduced, thus, generating reduced GO (r-GO). PE-alt-DA6 was prepared by melt mixing hexamethylenediamine (HMDA) and polyethylene-alt-maleic anhydride (PE-alt-MA). The mechanical, thermal, rheological and morphological properties of the PP/GO nanocomposites were examined by keeping the compatibilizer content constant, while varying the amount of GO in the polymer matrix (0.5, 1.0, 3.0 and 5.0 wt %).

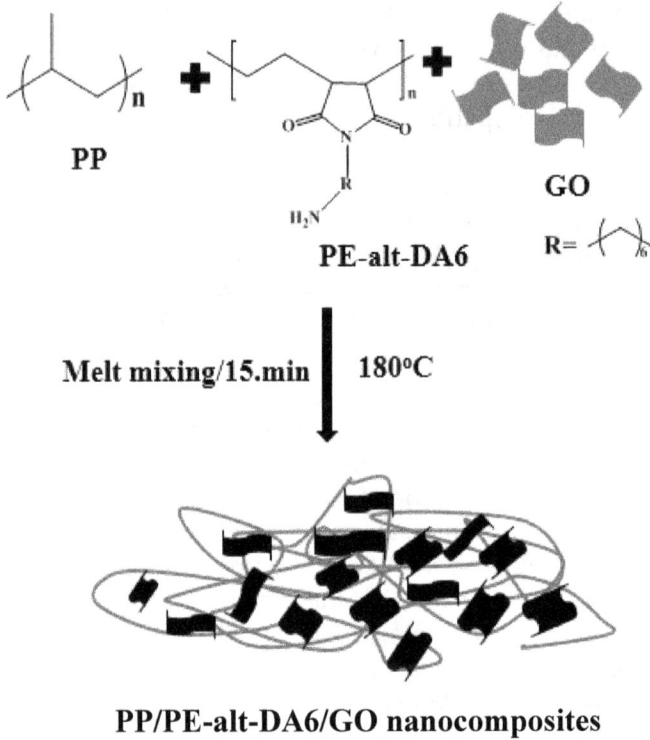

PP/PE-alt-DA6/GO nanocomposites

Figure 1.1 Generation of PP/GO nanocomposites from PP, PE-alt-DA6 and GO.

1.2 Experimental

1.2.1 Materials

The isotactic PP used in this study was supplied by Abu Dhabi Polymers Company (Borouge), with melt flow rate (MFR) of 9 g/min (200 °C, 2.16 kg). HMDA, dimethyl formamide (DMF) and PE-alt-MA were purchased from Sigma Aldrich, Germany. PE-alt-MA consisted of ≤0.2 wt% maleic anhydride and ≤1.5 wt% water. It had an average molecular weight in the range of 100,000-500,000 g/mol, viscosity in the range of 0.75- 1.05 poise and glass transition temperature of about 235 °C. HMDA had a melting point in the range of 42-45 °C. GO nanoplatelets were generated by the oxidation of graphite, as reported earlier [27]. The polymers and reagents were used as received without any further purification.

1.2.2 Generation of PE-alt-DA6

To prepare PE-alt-DA6 compatibilizer, 50 g PE-alt-MA and 10 g HMDA were taken in a round bottomed flask, followed by the immersion of the flask in a heating oil bath at 170 °C under reflux and stirring. At the top of the reflex condenser, a guard tube was placed to prevent the moisture entering the flask. The mixture turned yellow after the completion of the reaction. After cooling to the room temperature, distilled water (200 mL) was added to the flask. After overnight stirring, the mixture was added to 1.5 L methanol and stirred for 2 h. Subsequently, the filtered residue was added to 500 mL of 8% NaHCO₃ solution and re-filtered. Further, the residue was washed with water, followed by washing with ethanol. The product was finally dried at of 50 °C for 12 h. The chemical reaction describing the generation of PE-alt-DA6 is shown in Figure 1.2.

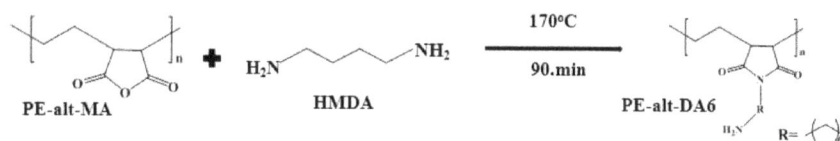

Figure 1.2 Generation of PE-alt-DA6.

1.2.3 Generation of Compatibilizer/Filler Hybrid

For identifying the chemical reaction between PE-alt-DA6 and GO, their hybrid was prepared without PP. Here, GO (0.2 g) was added to 100 mL DMF solvent, followed by ultra-sonication to achieve effective GO dispersion. PE-alt-DA6 (0.2 g) was subsequently added to the dispersion and stirred for 5 h at 110 °C. Later, centrifugation and separation of the solvent were carried out. Finally, the hybrid PE-alt-DA6/GO was obtained, which was examined using Fourier transform infrared spectroscopy (FTIR) spectroscopy to analyze the chemical bonding between PE-alt-DA6 and GO.

1.2.4 Preparation of Nanocomposites

The PP nanocomposites were generated by melt mixing PP, GO and PE-alt-DA6, utilizing a twin screw extruder Thermo Scientific Haake Minilab II. The extruder was operated in co-rotating mode at 100 rpm and 180 °C. The materials were melt mixed for about 15 min. Table 1.1 also depicts the composite formulations. In all combinations, the amount of PE-alt-DA6 compatibilizer was fixed at 10 wt%. After extrusion, the test specimens were injection molded using Thermo Scientific Haake Minijet Pro molding machine. The mold and cylinder temperatures were 135 °C and 190 °C, respectively. Also, the injection and post pressures were 430 and 500 bar, respectively, applied for 10 s.

Table 1.1 Composition of PP nanocomposites (in wt%)

Polymer/ Nanocomposites	PP (%)	GO (%)	PE-alt-DA6 (%)
Pure PP	100	0.0	0
PP/PE-alt-DA6	90.0	0.0	10
PP/PE-alt-DA6/0.5% GO	89.5	0.5	10
PP/PE-alt-DA6/1.0% GO	89.0	1.0	10
PP/PE-alt-DA6/3.0% GO	87.0	3.0	10
PP/PE-alt-DA6/5.0% GO	85.0	5.0	10

1.2.5 Characterization

FTIR was performed on a Nicolet iS10 spectrometer having SmartiTR diamond ATR accessory, DTGS KBr detector and KBr beam splitter. The analysis was carried out in transmission mode. Thermogravimetric analysis (TGA) was performed on a Discovery thermogravimetric analyzer from TA Instruments over the temperature range of 35 °C to 700 °C, in an inert nitrogen atmosphere, using a heating rate of 10 °C/min. Differential scanning calorimetry (DSC) analysis of the nanocomposites was carried out using a Discovery DSC from TA Instruments. Sample weight was approx.. 5-8 mg, and nitrogen was employed as the carrier gas with a constant flow rate. The samples were subjected to two cycles of heating and cooling at a rate of 10 °C/min. The specimens were heated from room temperature (25 °C) to 200 °C, followed by cooling to -60 °C. Subsequently, the samples were heated to 200 °C and then cooled down to 25 °C.

Tensile strength, tensile modulus and elongation at break of the samples were measured using Instron 5567 UTM at the strain rate of 10 mm/min. The tests were performed at room temperature in accordance with ASTM D638. The Izod impact strength analysis was performed on rectangular bar shaped specimens according to ASTM

D256 at room temperature. An average value from five test specimens was recorded. The melt rheological performance of pure PP and its nanocomposites was analyzed using a AR2000 813901 rheometer from TA Instruments. The analysis was performed using an angular frequency range of 0.1-500 rad/sec, at 2% strain and 180 °C. The Raman spectroscopic analysis was performed using Horiba LabRAM HR, equipped with 633 nm laser source.

Microscopy analysis of the nanocomposite samples was carried out using FEI, TECNAI transmission electron microscope (TEM) at an acceleration voltage of 200 kV. The sample sections were microtomed using PowerTome microtome, equipped with a diamond knife. For image processing, digital micrograph software (Gatan, USA) was employed. Panalytical X'Pert Pro diffractometer was used to analyze the wide angle X-ray diffraction (WAXRD) of materials in reflection mode, and the scanning angle range was 2θ = 5-60°. Laser flash analysis (LFA) was carried out using Netzsch LFA 447 to determine the thermal conductivity of PP nanocomposites.

1.3 Results and Discussion

FTIR analysis was performed to determine the nature of interactions between GO, PP and PE-alt-DA6 compatibilizer. Figure 1.3 presents the FTIR spectra of pure PP, GO, PE-alt-DA6, PE-alt-DA6/GO hybrid and PP based nanocomposites. Pure PP exhibited distinctive peaks at 1167 cm^{-1} (C-C stretching of C-CH$_3$ group), 1375 cm^{-1} (symmetric C-H bending vibration of –CH$_3$ and –CH$_2$ groups), 1455 cm^{-1} (asymmetric C-H bending vibration of –CH$_3$ and –CH$_2$ groups), 2837 cm^{-1} (C-H symmetric stretching of –CH$_2$ group), 2866 cm^{-1} (C-H symmetric stretching of –CH$_3$ group), 2916 cm^{-1} (C-H asymmetric stretching of –CH$_2$ group) and 2949 cm^{-1} (C-H asymmetric stretching of –CH$_3$ group). For pristine GO, the FTIR spectrum confirmed the presence of characteristic oxygen containing groups. Specifically, the peaks at 1225 cm^{-1} (C-O-C stretching vibration), 1393 cm^{-1} (O-H deformation vibration), 1625 cm^{-1} (C=C stretching vibration), 1733 cm^{-1} (C=O stretching vibration) and 3222 cm^{-1} (O-H stretching vibration) were observed [28]. For PE-alt-DA6 compatibilizer, specific peaks were observed at 1686 cm^{-1} (1° amide C=O stretching), 1438 cm^{-1} (N-H secondary amine stretching) and 1402 cm^{-1} (amide carbonyl stretching), among others. Further, in the case of PE-alt-DA6/GO hybrid, a shift of the aromatic C=C stretching vibration was observed from 1625 cm^{-1} to 1655 cm^{-1}, due to r-GO formation. Thus, the FTIR analysis confirmed the effective bond formation between PE-alt-DA6 and GO, to form r-GO.

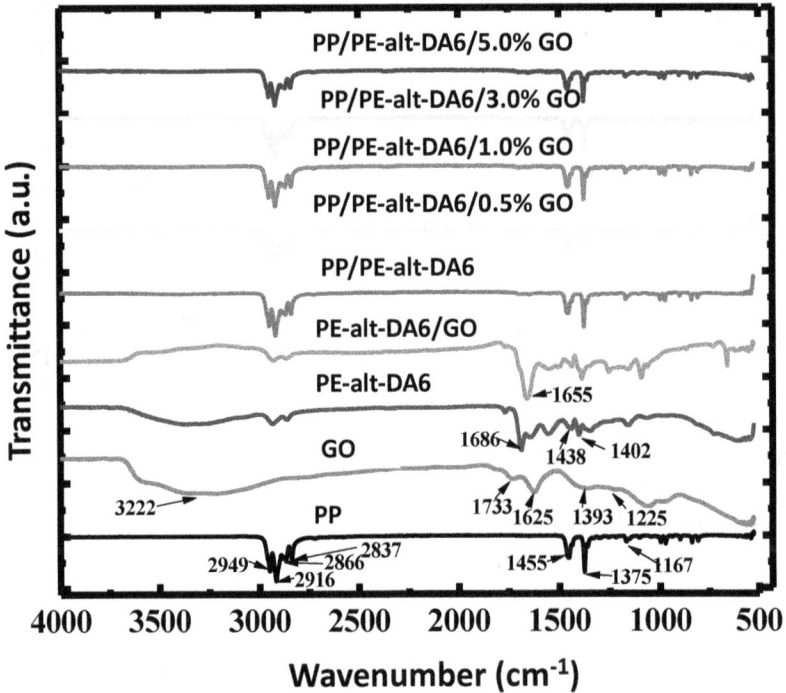

Figure 1.3 FTIR spectra of pristine PP, GO, PE-alt-DA6, PE-alt-DA6/GO hybrid and PP/PE-alt-DA6/GO nanocomposites.

Raman spectroscopic analysis was performed to gain further insights about the interaction between the GO and PE-alt-DA6 compatibilizer phases (Figure 1.4). The dominant feature of the GO spectrum is the prominent D and G peaks observed at almost 1352 and 1577 cm^{-1}, respectively [29], where the G peak indicates the graphitic carbon structure [30]. Using the G band, the thickness of graphene layers has also been estimated. The D peak provides information about the graphene disorder arising because of the structural edge effects [31,32]. From the Raman spectra of GO and PE-alt-DA6/GO hybrid, the D band of the specimens was observed to remain unchanged. However, due to the covalent bonding between PE-alt-DA6 and GO, the G band of the hybrid shifted from 1577 cm^{-1} to 1595 cm^{-1}. This indicated the reduction of GO to form r-GO. The evaluation of I_D/I_G ratio (I_D: Raman intensity of D peak and I_G: Raman intensity of G peak) of GO and PE-alt-DA6/GO hybrid is also presented in Table 1.2 [33]. When compared to GO, the I_D/I_G ratio of the hybrid PE-alt-DA6/GO was enhanced by 10%. The removal of the oxygen containing groups in GO nanofiller would have resulted in the observed enhancement. Additionally, it lowered the average size of the sp^2 domain.

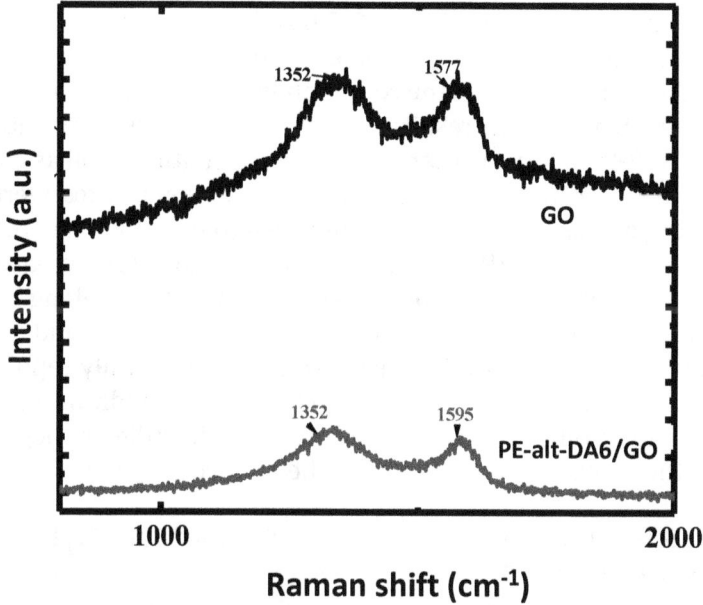

Figure 1.4 Raman spectra of pristine GO and PE-alt-DA6/GO hybrid.

Table 1.2 Raman intensity ratio (I_D/I_G) of pristine GO and PE-alt-DA6/GO hybrid

Material	D band (cm⁻¹)	G band(cm⁻¹)	Raman intensity ratio (I_D/I_G)
GO	1352	1577	1.00
PE-alt-DA6/GO hybrid	1352	1595	1.10

It has been observed that the mechanical properties of the nano-composites depend on filler dispersion state, percentage crystallinity and polymer-filler interactions [34-36]. Wypych [37] also stated that the polymer-filler interactions are a critical factor influencing the tensile properties of a polymer-filler system. To gain qualitative insights about this aspect, Table 1.3 illustrates the tensile strength, modulus and elongation at break of pure PP, PP/PE-alt-DA6 hybrid and PP/PE-alt-DA6/GO nanocomposites. Figure 1.5 also presents the tensile modulus of the PP/PE-alt-DA6/GO nanocomposites as a function of GO content. A significant increase in the tensile modulus was observed in the composites as compared to pure PP. For instance, the tensile modulus of PP/PE-alt-DA6/5.0% GO increased by 88% with respect to pure PP. Such improvement can be attributed to the reaction of the amine groups in the compatibilizer with the oxy-functional graphene sheets, which resulted in the homogenous distribution of GO within PP [38,39]. Further, the tensile strength as well as elongation at break of the PP/PE-alt-DA6/GO nanocomposites were observed to decrease marginally as compared to pure PP. The presence of alt groups in the PP/PE-alt-DA6/GO nanocomposites increased the crosslink density of the samples, thereby, reducing the elongation at break [40]. The marginal reduction in the yield strength and elongation was in contrast with the literature studies generally reporting a significant reduction in these properties on the addition of nanofillers. In addition, the addition of even 5 wt% filler did not negatively impact the mechanical properties of the nanocomposites.

Table 1.3 Tensile properties of pure PP, PP/PE-alt-DA6 and PP/PE-alt-DA6/GO nanocomposites

Material	Elongation at break (%)	Yield stress (MPa)	Tensile modulus (MPa)
Pure PP	7	38	879
PP/PE-alt-DA6	5	34	1447
PP/PE-alt-DA6/0.5% GO	5	34	1509
PP/PE-alt-DA6/1.0% GO	5	35	1547
PP/PE-alt-DA6/3.0% GO	5	35	1632
PP/PE-alt-DA6/5.0% GO	5	34	1654

The fracture toughness of a material represents the amount of energy absorbed for fracturing the sample. The improvement in the toughness of PP due to the addition of the nanoparticles is mainly because of the micro-cracking mechanism [41]. The addition of fillers in the polymer results in the establishment of a heterogeneous system [42]. Hence, an external load can generate stress concentration in the polymer composites, thereby influencing the fracture behavior of the material [43]. The fracture in the polymer starts with the plastic deformation before the initial crack. Thus, the debonding of particles occurs, and the zones are stretched and ruptured [44,45].

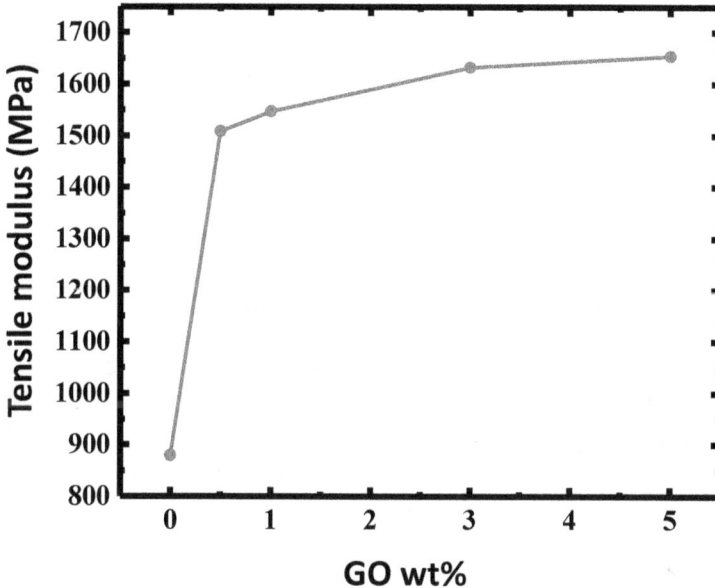

Figure 1.5 Tensile modulus of PP/PE-alt-DA6/GO nanocomposites as a function of GO content.

The findings from the impact analysis of pristine PP, PP/PE-alt-DA6 and PP/PE-alt-DA6/GO nanocomposites are presented in Table 1.4. The impact strength of pure PP was observed to be 7.96 kJ/m². The impact strength of PP/PE-alt-DA6 was observed to increase by 18% as compared to pure PP. For PP/PE-alt-DA6/GO nanocomposites, the addition of 0.5 wt%, 1.0 wt% and 3.0 wt% GO enhanced the impact strength to 11.06 kJ/m² (39% increase), 11.48 kJ/m² (44% increase) and 13.52 kJ/m² (70% increase) respectively. Further increase in GO content resulted in a reduction in the impact strength of the composite. This was due to the greater stacking of the graphene sheets within the PP matrix [46,47], caused due to van der Waals forces. The results also indicated that the PE-alt-DA6 compatibilizer led to substantial bonding between the GO and PP phases, which delayed the crack initiation as well as propagation in the PP nanocomposites. The enhancement in the crack growth initiation energy, thus, caused a significant increase in the toughness of the PP nanocomposites [46]. Overall, a remarkable increase in the impact strength as well as a good toughness-stiffness balance was observed for PP/PE-alt-DA6/3.0% GO.

Table 1.4 Impact strength of PP/PE-alt-DA6/GO nanocomposites

Material	Unnotched Izod impact strength (kJ/m²)
Pure PP	7.96
PP/PE-alt-DA6	9.41
PP/PE-alt-DA6/0.5% GO	11.06
PP/PE-alt-DA6/1.0% GO	11.48
PP/PE-alt-DA6/3.0% GO	13.52
PP/PE-alt-DA6/5.0% GO	10.34

The rheological analysis is performed to evaluate the filler dispersion, internal structure and processing properties of nanocomposites [48]. Figures 1.6-1.9 demonstrate the storage (G') and loss modulus (G"), mechanical loss factor (tan δ) and complex viscosity (η*) of the PP composites as a function of angular frequency (ω). The G' of the PP/PE-alt-DA6/GO nanocomposites was observed to be lower than pristine PP. The reduced G' value of the nanocomposites could be attributed to the matrix plasticization [49]. Due to the low surface friction of GO, the PP/GO interlayer slipping occurs. Therefore, the PP nanocomposite melt behaves like a viscous PP liquid [50]. Additionally, the G' curves were observed to converge with each other at higher frequencies. This clearly indicated the concentration independence behavior of the materials at such frequencies. G" of pristine PP was higher than G' at lower angular frequency values indicating the dominance of the viscous behavior.

Tan δ of PP/PE-alt-DA6/GO nanocomposites was observed to be higher than pristine PP. It was noticed from the Figure 1.8 that the pristine PP and PP/PE-alt-DA6/GO nanocomposites showed the viscoelastic nature at tan δ value of about 0.2 rad/sec, and glassy nature at almost 100 rad/sec [51,52].

Newtonian fluid is represented by a horizontal line in the η* vs. ω plot. At the same time, the shear thinning of the sample is indicated by a reduction in the complex viscosity with increase in ω or shear rate [53]. In PP nanocomposites too, η* was observed to decrease with increase in ω (Figure 1.9), revealing the shear thinning nature. The shear thinning behavior at higher ω is related to the disentaglement mechanism. In addition, η* of PP nanocomposites was observed to be lower than pristine PP, especially at lower frequencies.

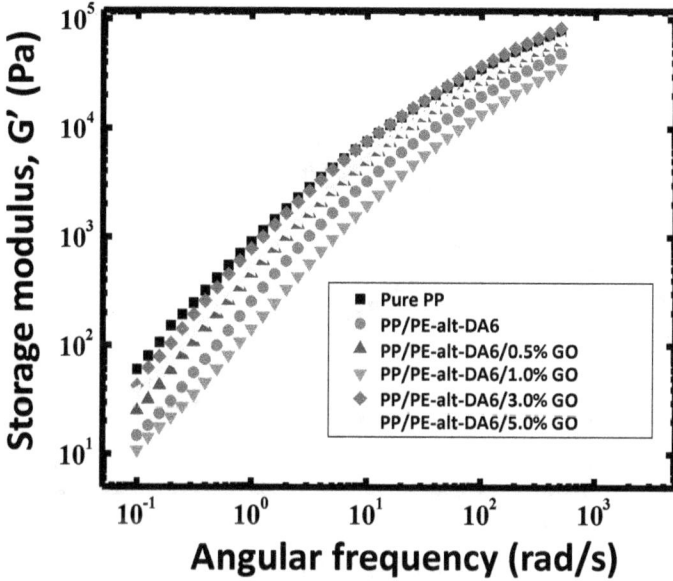

Figure 1.6 Storage modulus (G') vs. angular frequency for pristine PP, PP/PE-alt-DA6 blend and PP/PE-alt-DA6/GO nanocomposites.

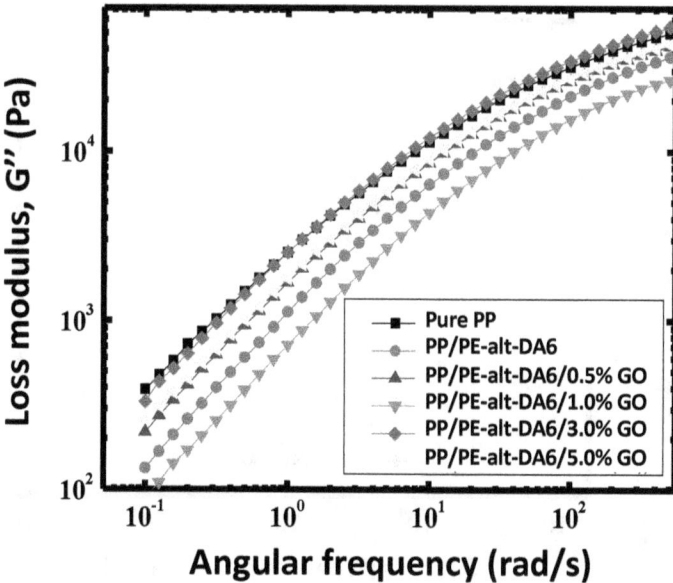

Figure 1.7 Loss modulus (G") vs. angular frequency for pristine PP, PP/PE-alt-DA6 blend and PP/PE-alt-DA6/GO nanocomposites.

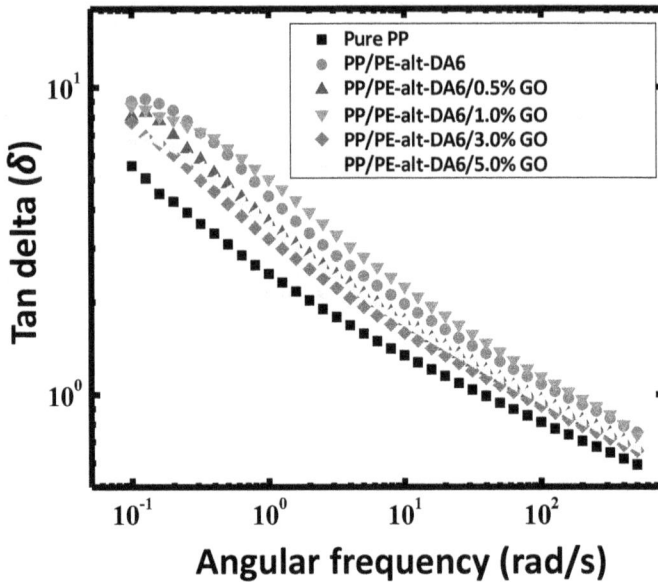

Figure 1.8 Tan delta vs. angular frequency for pristine PP, PP/PE-alt-DA6 blend and PP/PE-alt-DA6/GO nanocomposites.

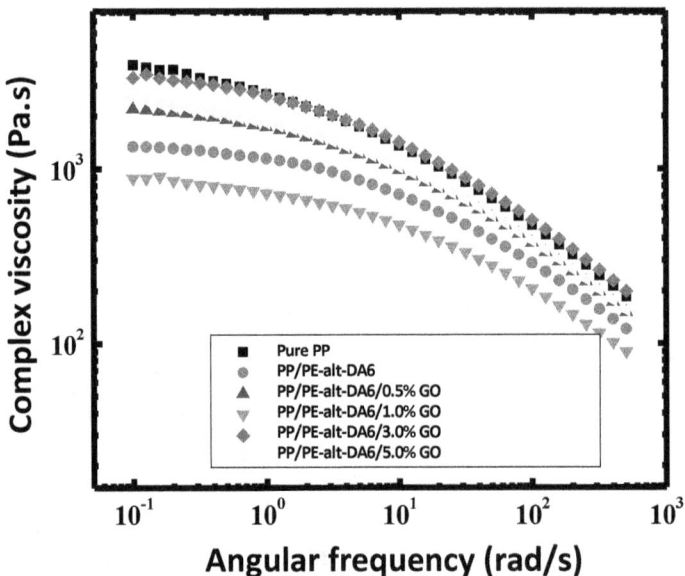

Figure 1.9 Complex viscosity (η^*) vs. angular frequency for pristine PP, PP/PE-alt-DA6 blend and PP/PE-alt-DA6/GO nanocomposites.

The calorimetric properties of pristine PP and PP/PE-alt-DA6/GO nanocomposites are presented in Table 1.5. The Figures 1.10 and 1.11 also illustrate the melting and cooling thermograms of the materials. The degree of crystallinity was calculated using the enthalpy of fusion of 100% crystalline PP (209 J/g) [54].

Table 1.5 Calorimetric properties of pristine PP and PP/PE-alt-DA6/GO nanocomposites

	Heating			Cooling			X_c
	ΔH_m (cal/g)	ΔH_m (J/g)	T_m (°C)	ΔH_c (cal/g)	ΔH_c (J/g)	T_c (°C)	
Pure PP	20.8	86.9	167	22.7	95.1	128	42
PP/PE-alt-DA6/0.5% GO	21.3	89.1	167	22.0	92.2	127	43
PP/PE-alt-DA6/1.0% GO	21.3	89.1	168	21.7	90.9	126	43
PP/PE-alt-DA6/3.0% GO	21.5	89.9	169	21.3	89.2	126	44
PP/PE-alt-DA6/5.0% GO	21.2	88.7	168	20.3	84.8	126	45

A single endothermic peak was observed in the melting thermograms of all samples at temperature >160 °C. This indicated the presence of the α- crystals solely, with no β- crystals [55]. The incorporation of PE-alt-DA6 and GO to the PP matrix marginally enhanced the crystallinity, which was also a function of GO fraction. The observed improvement in the crystallinity can be related to the heterogeneous nucleation effect in the PP matrix. Additionally, no remarkable change in the T_m and T_c values of PP was observed on the incorporation of PE-alt-DA6 and GO. Several literature studies have reported that the grafting of long chain alkylamines, such as decylamine (A10), hexadecylamine (A16) and octadecylamine (A18), on the 2-D nanosheets resulted in the oriented crystalline arrangement of the alkylamines [56]. However, the compatibilizer used in the study was largely amorphous in nature. Thus, the observed changes in crystallinity in the nanocomposites resulted from the combined effects of the compatibilizer and GO.

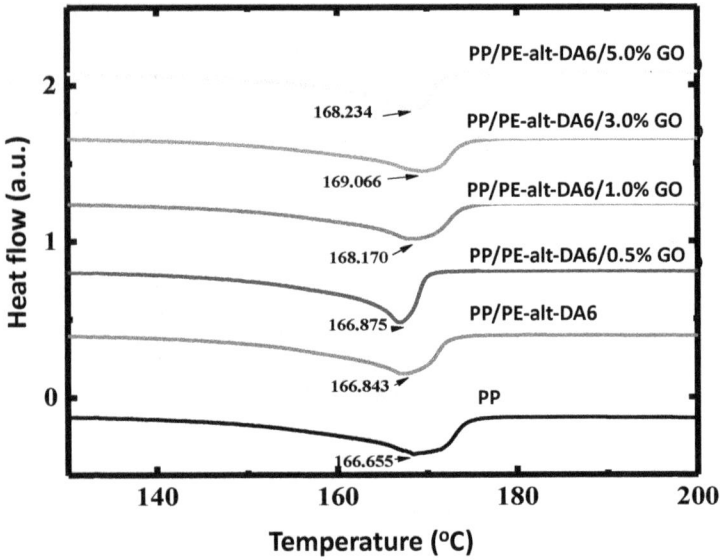

Figure 1.10 Melting endotherms of pristine PP, PP/PE-alt-DA6 blend and PP/PE-alt-DA6/GO nanocomposites.

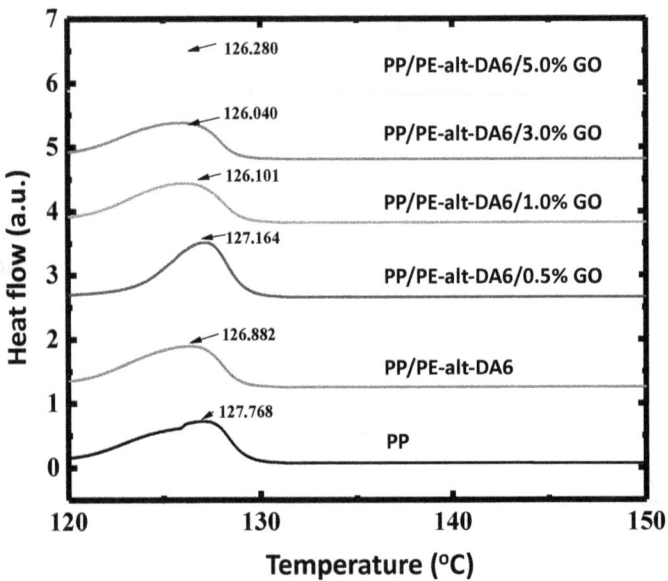

Figure 1.11 Crystallization isotherms of pristine PP, PP/PE-alt-DA6 blend and PP/ PE-alt-DA6/GO nanocomposites.

Figure 1.12 presents the TGA thermograms of pristine PP, PP/PE-alt-DA6 blend and PP/PE-alt-DA6/GO nanocomposites. A single step degradation mechanism was observed for all the materials. The T_{ini} (5% weight loss temperature), $T_{40\%}$ (40% weight loss temperature) and char yield (mass remaining at 700 °C) are also compiled in Table 1.6.

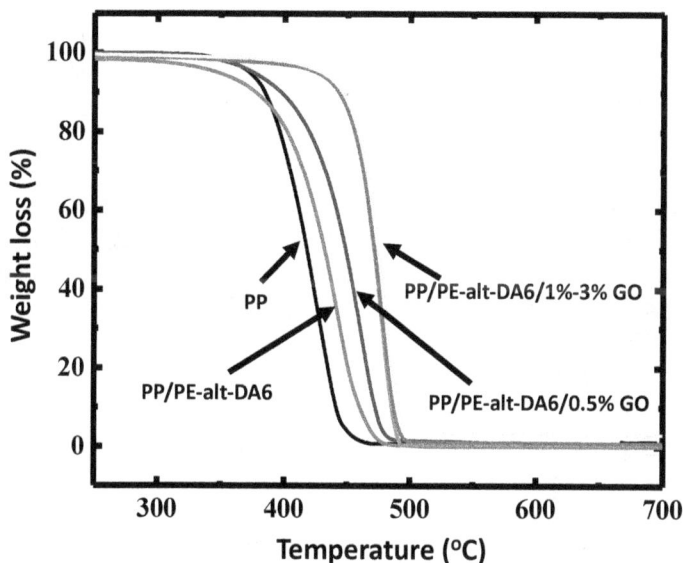

Figure 1.12 TGA thermograms of pristine PP, PP/PE-alt-DA6 blend and PP/PE-alt-DA6/GO nanocomposites

The addition of PE-alt-DA6 and GO nanofiller enhanced the thermal stability to a large extent (except the initiation of degradation temperature for the PP/PE-alt-DA6 blend). The thermal degradation temperatures for 5% as well as 40% weight loss increased with the GO loading until 1.0 wt%. The degradation temperatures were observed to be similar for the PP nanocomposites having GO content higher than 1.0 wt%. The observed thermal behavior revealed high thermal stability of the PP/PE-alt-DA6/GO nanocomposites, as compared to pristine PP [57]. The char yield of PP/PE-alt-DA6/GO nanocomposites also increased with GO content, as compared to pure PP. Overall, the enhancement in the thermal stability of the PP nanocomposites was due to the stronger interactions between the GO and PP phases, which enhanced the heat transfer at the interface, thus, increasing the thermal degradation activation energy [58].

Table 1.6 Thermal degradation behavior of pristine PP, PP/PE-alt-DA6 blend and PP/PE-alt-DA6/GO nanocomposites

	$T_{ini\%}$ (°C)	$T_{40\%}$ (°C)	Char yield at 700 °C (wt %)
Pure PP	375	414	1.26
PP/PE-alt-DA6	347	424	0.00
PP/PE-alt-DA6/0.5% GO	375	441	0.53
PP/PE-alt-DA6/1.0% GO	425	467	0.58
PP/PE-alt-DA6/3.0% GO	428	467	0.92
PP/PE-alt-DA6/5.0% GO	430	470	3.04

The thermal conductivity analysis of PP/PE-alt-DA6/GO nano-composites is presented in in Figure 1.13. The incorporation of reduced GO enhanced the thermal conductivity of PP significantly. For instance, in PP/PE-alt-DA6/5.0% GO composite, the thermal conductivity was observed to be 0.336 W/mK, which was 68% higher than pristine PP [59]. The pristine graphene has a thermal conductivity >5000 W/mK [60], however, a much lower degree of increment was observed in the nanocomposites owing to various factors such as structural defects in the filler, incomplete reduction of GO and GO aggregation in the composites.

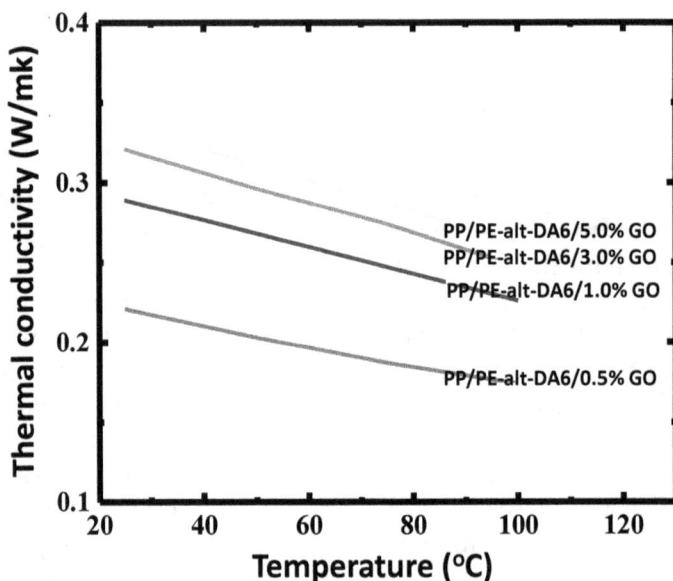

Figure 1.13 Thermal conductivity of PP/PE-alt-DA6/GO nanocomposites.

The physical and chemical properties of the nanocomposites are mainly driven by the polymer-filler interactions as well as filler dispersion [61]. Figure 1.14 presents the morphological analysis of the PP/PE-alt-DA6/5.0% GO composite to gain insights about the extent of filler dispersion. The composite exhibited well dispersed and exfoliated morphology, thus, also indicating optimal polymer-filler interactions [34].

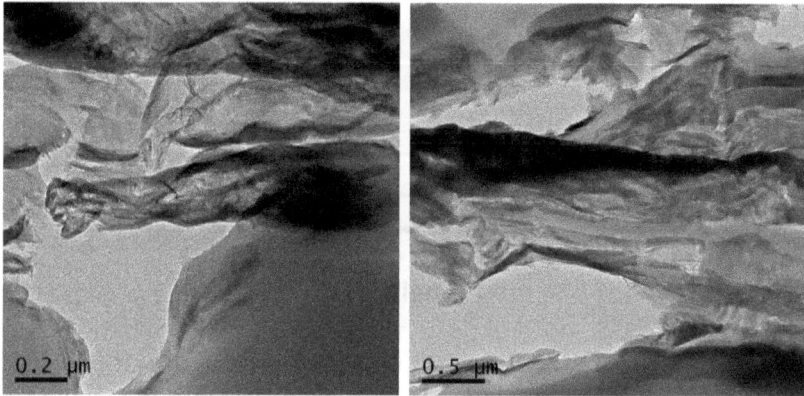

Figure 1.14 TEM images of PP/PE-alt-DA6/5.0% GO nanocomposite.

Figure 1.15 presents the WAXRD patterns of pristine PP, GO, PE-alt-DA6, PP/PE-alt-DA6 blend and PP/PE-alt-DA6/GO nanocomposites. In the case of pristine GO, a strong reflection peak was observed at 2θ = 9.6°. The diffraction peaks of pure PP were observed at 14.5°, 17.2°, 18.8° as well as 22.1°, which corresponded to the (110), (040), (130) and (111) planes of the PP crystal, respectively [62]. The XRD pattern of PE-alt-DA6 exhibited no sharp peak, suggesting the amorphous nature. Broad and weak diffraction peaks were noticed in the diffraction patterns of PP/PE-alt-DA6/GO nanocomposites at 2θ range 25-30°, due to the reduction of GO [63]. As previously identified in the DSC analysis, no β and γ phases were noticed in the materials.

1.4 Conclusion

Melt mixing method was used to generate PP/PE-alt-DA6/GO nanocomposites, by varying the content of GO (0.5, 1.0, 3.0 and 5.0 wt%) and keeping the concentration of PE-alt-DA6 compatibilizer constant (10.0 wt%). The microscopy analysis confirmed the homogenous dispersion of GO in the PP matrix owing to the presence of PE-alt-DA6

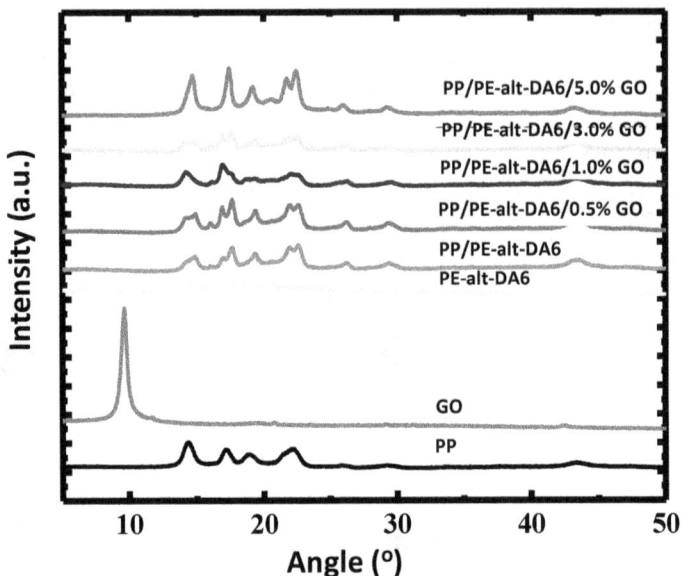

Figure 1.15 XRD patterns of pure PP, GO, PE-alt-DA6, PP/PE-alt-DA6 and PP/PE-alt-DA6/GO nanocomposites.

compatibilizer. Additionally, the formation of a bond between GO and PE-alt-DA6, forming r-GO in the PP nanocomposites, was confirmed by Raman, FTIR and XRD analyses. The effective polymer-filler inter-actions resulted in an enhancement in the mechanical and thermal properties of the resultant PP nanocomposites. These also delayed the crack initiation and propagation in the PP/PE-alt-DA6/GO nano-composites, thereby increasing the toughness of the nanocomposites. For instance, PP/PE-alt-DA6/3.0% GO exhibited a 70% increase in toughness as compared to pure PP. Shear thinning nature of the sam-ples was noted at higher frequencies in the rheological analysis. The DSC analysis indicated that the crystallinity of PP enhanced to a small extent in the nanocomposites. The thermal conductivity of PP was also observed to increase by the addition of GO. A 68% increase in the thermal conductivity of PP/PE-alt-DA6/5.0% GO was observed, as compared to pristine PP.

References

1. Pasquini, N., and Sgarzi, P. (2005) Polypropylene - The market. In: *Polypropylene Handbook*, Pasquini, N. (ed.), 2nd edition, Carl Hanser

Verlag, Germany.

2. Kissel, W. J., Han, J. H., Meyer, J. A. (2003) Polypropylene: Structure, properties, manufacturing and applications. In: *Handbook of Polypropylene*, Karian, H. (ed.), Marcel Dekker AG, Switzerland, pp. 10-27.

3. Lin, Y., Chen, H., Chan, C. M., Wu, J. (2008) High impact toughness polypropylene/CaCO₃ nanocomposites and the toughening mechanism. *Macromolecules*, **41**, 9204-9213.

4. Lopez-Quintanilla, L. M., Sanchez-Valdes, S., Ramos-de-Valle, L. F., and Miranda, R. G. (2006) Preparation and mechanical properties of PP/PP-g-MA/Org-MMT nanocomposites with different MA content. *Polymer Bulletin*, **57**, 385-393.

5. *Advances in Polyolefin Nanocomposites*, Mittal, V. (ed.), CRC Press, USA (2011).

6. *Polymer-Graphene Nanocomposites*, Mittal, V. (ed.), Royal Society of Chemistry, UK (2012).

7. Jancar, J., and Di Benedetto, A. T. (1994) The mechanical properties of ternary composites of polypropylene with inorganic fillers and elastomer inclusions. *Journal of Material Science*, **29**, 4651-4658.

8. Arencon, D., Velasco, J. I., Realinho, V., Sanchez-Soto, M. A., and Gordillo, A. (2007) Fracture toughness of glass microsphere filled polypropylene and polypropylene/poly(ethylene terephthalate-coisophthalate) blend-matrix composites. *Journal of Material Science*, **42**, 19-29.

9. Hong, J.-Y., Shin, K.-Y., Kwon, O. S., Kang, H., and Jang, J. (2011) A strategy for fabricating single layer graphene sheets based on a layer-by-layer self assembly. *Chemical Communications*, **47**, 7182-7184.

10. Kuilla, T., Bhadra, S., Yao, D., Kim, N. H., Bose, S., and Lee, J. H. (2010) Recent advances in graphene based polymer composites. *Progress in Polymer Science*, **35**(11), 1350-1375.

11. Song, P., Cao, Z., Cai, Y., Zhao, L., Fang, Z., and Fu, S. (2011) Fabrication of exfoliated graphene-based polypropylene nanocomposites with enhanced mechanical and thermal properties. *Polymer*, **52**(18), 4001-4010.

12. Varghese, A. M., Rangaraj, V. M., Mun, S. K., Macosko, C. W., and Mittal, V. (2018) Effect of graphene on polypropylene/maleic anhydride-graft-ethylene-vinyl acetate (PP/EVA-g-MA) blend: Mechanical, thermal, morphological, and rheological properties. *Industrial & Engineering Chemistry Research*, **57**, 7834-7845.

13. Li, D., Muller, M. B., Gilje, S., Kaner, R. B., and Wallace, G. G. (2008) Processable aqueous dispersions of graphene nanosheets. *Nature Nanotechnology*, **3**, 101-105.

14. Layek, R. K., and Nandi, A. K. (2013) A review on synthesis and properties of polymer functionalized graphene. *Polymer*, **54**(19), 5087-

5103.

15. Cao, Y., Feng, J., and Wu, P. (2012) Polypropylene-grafted graphene oxide sheets as multifunctional compatibilizers for polyolefin-based polymer blends. *Journal of Materials Chemistry*, **22**, 14997-15005.

16. Szabo, T., Berkesi, O., and Dekany, I. (2005) DRIFT study of deuterium-exchanged graphite oxide. *Carbon*, **43**(15), 3186-3189.

17. Eda, G., and Chhowalla, M. (2010) Chemically derived graphene oxide: Towards large-area thin-film electronics and optoelectronics. *Advanced Materials*, **22**, 2392-2415.

18. Stankovich, S., Piner, R. D., Nguyen, S. T., and Ruoff, R. S. (2006) Synthesis and exfoliation of isocuanate-treated graphene oxide nanoplastelets. *Carbon*, **44**(15), 3342-3347.

19. Park, S., Dikin, D. A., Nguyen, S. T., and Ruoff, R. S. (2009) Graphene oxide sheets chemically cross-linked by polyallylamine. *Journal of Physical Chemistry C*, **113**(36), 15801-15804.

20. Shin, K.-Y., Hong, J.-Y., Lee, S. and Jang, J. (2012) Evaluation of anti-scratch properties of graphene oxide/polypropylene nanocomposites. *Journal of Materials Chemistry*, **22**, 7871-7879.

21. Kobayashi, S., Song, J., Silvis, H. C., Macosko, C. W., and Hillmyer, M. A. (2011) Amino-functionalized polyethylene for enhancing the adhesion between polyolefins and polyurethanes. *Industrial and Engineering Chemistry Research*, **50**(6), 3274-3279.

22. Zhou, X.-Q., and Hay, J. N. (1993) Structure property relationships in annealed blends of linear low density polyethylene with isotactic polypropylene. *Polymer*, **34**(22), 4710-4716.

23. Hill, M. J., Oiarzabal, L., and Higgins, J. S. (1994) Preliminary studies of polypropylene/ linear low density polyethylene blends by transmission electron microscopy. *Polymer*, **35**(15), 3332-3337.

24. Jose, S., Aprem, A. S., Francis, B., Chandy, M. C., Werner, P., Alstaedt, V., and Thomas, S. (2004) Phase morphology, crystallisation behaviour and mechanical properties of isotactic polypropylene/high density polyethylene blends. *European Polymer Journal*, **40**(9), 2105-2115.

25. Fel, E., Khrouz, L., Massardier, V., Cassagnau, P., and Bonneviot, L. (2016) Comparative study of gamma-irradiated PP and PE polyolefins part 2: Properties of PP/PE blends obtained by reactive processing with radicals obtained by high shear or gamma-irradiation. *Polymer*, **82**, 217-227.

26. Sánchez-Valdes, S., Méndez-Nonell, J., Medellín-Rodríguez, F. J., Ramírez-Vargas, E., Martínez-Colunga, J. G., de Valle, L. F. R., Mondragón-Chaparro, M., López-Quintanilla, M. L., García-Salazar, M. L. (2010) Evaluation of different amine-functionalized polyethylenes as compatibilizers for polyethylene film nanocomposites. *Polymer International*, **59**(5), 704-711.

27. Chaudhry, A. U., and Mittal, V. (2013) High density polyethylene

nanocomposites using master batches of chlorinated polyethylene/Graphene oxide. *Polymer Engineering and Science*, **53**(1), 78-88.

28. Tien, H. N., Luan, V. H., Lee, T. K., Kong, B. S., Chung, J. S., Kim, E. J., and Hur, S. H. (2012) Enhanced solvothermal reduction of graphene oxide in a mixed solution of sulfuric acid and organic solvent. *Chemical Engineering Journal*, **211-212**, 97-103.

29. Elias, D.C. Nair, R.R. Mohiuddin, T.M.G. Morozov, S.V. Blake, P. Halsall, M.P. Ferrari, A. C., Boukhvalov, D. W., Katsnelson, M. I., Geim, A. K., and Novoselov., K. S. (2009) Control of graphene's properties by reversible hydrogenation: Evidence for graphene. *Science*, **323**(5914), 610-613.

30. Lonkar, S. P., Pillai, V. V., Stephen, S., Abdala, A., and Mittal, V. (2016) Facile in situ fabrication of nanostructured graphene–CuO hybrid with hydrogen sulfide removal capacity. *Nano-Micro Letters*, **8**(4), 312-319.

31. Rao, C. N. R., Sood, A. K., Subrahmanyam, K. S., and Govindaraj, A. (2009) Graphene: the new two-dimensional nanomaterial. *Angewandte Chemie, International Edition*, **48** 7752-7777.

32. Xu, Z., Bando, Y., Liu, L., Wang, W., Bai, X., Golberg, D. (2011) Electrical conductivity, chemistry, and bonding alternations under graphene oxide to graphene transition as revealed by in situ TEM. *ACS Nano*, **5**(6), 4401-4406.

33. Jiao, L., Wang, X., Diankov, G., Wang, H., and Dai, H. (2010) Facile synthesis of high-quality graphene nanoribbons, *Nature Nanotechnology*, **5**, 321-325.

34. Ryu, S. H., and Shanmugharaj, A. M. (2014) Influence of long-chain alkylamine-modified graphene oxide on the crystallization, mechanical and electrical properties of isotactic polypropylene nanocomposites. *Chemical Engineering Journal*, **244**, 552-560.

35. Shen, B., Zhai, W., Tao, M., Lu, D., and Zheng, W. (2013) Chemical functionalization of graphene oxide toward the tailoring of the interface in polymer composites. *Composites Science and Technology*, **77**, 87-94.

36. Yuan, B., Bao, C., Song, L., Hong, N., Liew, K. M., and Hu, Y. (2014) Preparation of functionalized graphene oxide/polypropylene nanocomposites with significantly improved thermal stability and studies on the crystallization behavior and mechanical properties. *Chemical Engineering Journal*, **237**, 411-420.

37. Wypych G. (2016) *Handbook of Fillers*, 4th edition, Chem Tec Publishing, Canada.

38. Mandhakini, M., Chandramohan, A., Vengatesan, M. R., and Alagar, M. (2011) Synthesis and characterization of linseed vinyl ester fatty amide modified epoxy layered silicate Nanocomposites. *High Performance Polymers*, **23**(5), 403-412.

39. Qian, D., Dickey, E. C., Andrews, R., and Rantell, T. (2000) Load transfer and deformation mechanisms in carbon nanotube polystyrene composites. *Applied Physical Letters*, **76**, 2868.
40. Tamboli, S. M., Mhaske, S. T., and Kale, D. D. (2004) Crosslinked polyethylene. *Indian Journal of Chemical Technology*, **11**, 853-864.
41. *Polymer Toughening*, Arends, C. B. (ed.), Marcel Dekker Inc., USA (1996).
42. Kowalewski, T., Galeski, A., and Kryszewski, M. (1984) The structure and tensile properties of cold drawn modified chalk filled polypropylene. In: *Polymer Blends. Processing, Morphology and Properties*, Kryszewski, M., Galeski, A., and Martuscelli, E. (eds.), Plenum Press, USA, pp. 223-241.
43. Arencon, D., and Velasco, J. I., (2009) Fracture toughness of polypropylene-based particulate composites. *Materials*, **2**, 2046-2094.
44. Pukanszky, B., Vanes, M., Maurer, F. H. J., and Voros, G. (1994) Micromechanical deformations in particulate-filled thermoplastics: Volume strain measurements. *Journal of Material Science*, **29**, 2350-2358
45. Zhuk, A. V., Knunyants, N. N., Oshmyan, V. G., Topolkaraev, V. A., and Berlin, A. A. (1993) Debonding microprocesses and interfacial strength in particle-filled polymer materials. *Journal of Material Science*, **28**, 4995-5606.
46. Albdiry, M. T., Yousif, B. F., Ku, H., and Lau, K. T. (2012) A critical review on the manufacturing processes in relation to the properties of nanoclay/ polymer composites. *Journal of Composite Materials*, **47**(9), 1093-1115.
47. de Villoria, R. G., and Miravete, A. (2007) Mechanical model to evaluate the effect of the dispersion in nanocomposites. *Acta Materialia*, **55**, 3025-3031.
48. Woo, D. K., Kim, B. C. and Lee, S. J. (2009) Preparation and rheological behavior of polystyrene/multi-walled carbon nanotube composites by latex technology. *Korea-Australia Rheology Journal*, **21**(3), 185-191.
49. Chaudhry, A. U., and Mittal, V. (2014) Blends of high-density polyethylene with chlorinated polyethylene: Morphology, thermal, rheological, and mechanical properties. *Polymer Engineering and Science*, **54**, 85-95.
50. Ferry, J. D. (1980) *Viscoelastic Properties of Polymers*, 3rd edition, Wiley, New York.
51. Li, Y., Zhu, J., Wei, S., Ryu, J., Sun, L., and Guo, Z. (2011) Poly(propylene)/graphene nanoplatelet nanocomposites: melt rheological behavior and thermal, electrical, and electronic properties. *Journal of Macromolecular Chemistry and Physics*, **212**(18), 1951-1959
52. a) Chen, X., Wei, S., Yadav, A., Patil, R., Zhu, J., Ximenes, R., Sun, L., and Guo, Z. (2011) Poly(propylene)/carbon nanofiber nanocomposites:

Ex situ solvent-assisted preparation and analysis of electrical and electronic properties. *Macromolecular Materials and Engineering,* **296**, 434-443; b) Wei, S., Patil, R., Sun, L., Haldolaarachchige, N., Chen, X., Young, D. P., and Guo, Z. (2011) Ex situ solvent-assisted preparation of magnetic poly(propylene) 8nocomposites filled with Fe@ FeO Nanoparticles . *Macromolecular Materials and Engineering,* **296**(9), 850-857.

53. Sugimoto, M., Tanaka, T., Masubuchi, Y., Takimoto, J. I., and Koyama, K. (1999) Effect of chain structure on the melt rheology of modified polypropylene. *Journal of Applied Polymer Science,* **73**, 1493-1500.

54. *Polymer Handbook,* Brandup, J., and Immergut, E. H. (eds.), Wiley, USA (1988).

55. Bao, R.-Y., Cao, J., Liu, Z.-Y., Yang, W., Xie, B.-H., and Yang, M.-B. (2014) Towards balanced strength and toughness improvement of isotactic polypropylene nanocomposites by surface functionalized-graphene oxide. *Journal of Materials Chemistry A,* **2**(9), 3190-3199.

56. Lin, Z., Liu, Y., and Wong, C. P. (2010) Facile fabrication of superhydrophobic octadecylamine-functionalized graphite oxide film. *Langmuir,* **26**, 16110-16114.

57. Vengatesan, M. R., Singh, S., Pillai, V. V., and Mittal, V. (2016) Crystallization, mechanical, and fracture behavior of mullite fiber-reinforced polypropylene nanocomposites. *Journal of Applied Polymer Science,* **133**(30), doi: 10.1002/app.43725.

58. Li, B., and Zhong, W.-H. (2011) Review on polymer/graphite nanoplatelet nanocomposites. *Journal of Material Science,* **46**, 5595-5614.

59. Kalaitzidou, K., Fukushima, H., and Drzal, L. T. (2007) Multifunctional polypropylene composites produced by incorporation of exfoliated graphite nanoplatelets, *Carbon,* **45**, 1446-1452.

60. Prasher, R. (2010) Graphene spreads the heat. *Science,* **328**, 185-186.

61. Yang, L., Phua, S. L., Teo, J. K., Toh, C. L., Lau, S. K., Ma, J., and Lu, X. (2011) A biomimetic approach to enhancing interfacial interactions: polydopamine-coated clay as reinforcement for epoxy resin. *ACS Applied Materials & Interfaces,* **3**, 3026-3032.

62. Nishino, T., Matsumoto, T., and Nakamae, K. (2000) Surface structure of isotactic polypropylene by X-ray diffraction. *Polymer Engineering and Science,* **40**, 336-343.

63. Cao, N., and Zhang, Y. (2015) Study of reduced graphene oxide preparation by hummers' method and related characterization. *Journal of Nanomaterials,* **2015**, Article ID 168125.

2

Polypropylene-Sepiolite/Graphene Hybrid Composites: Thermal, Mechanical and Morphological Properties

2.1 Introduction

Layered silicates (clay) are one of the commonly used low cost nanofillers for polymer nanocomposites due to high surface area and aspect ratio, resulting in remarkable mechanical performance and thermal stability [1-4]. Different structural varieties of layered silicates, such as montmorillonite (MMT), bentonite, hectorite and sepiolite, have been used as nanofillers for the development of polymer nanocomposites [5-8]. More specifically, sepiolite is one of the naturally occurring clays and exhibits good textural properties with fibrous morphology. The crystalline structure of sepiolite clay consists of an octahedral sheet of magnesium hydroxide sandwiched between two tetrahedral sheets of silica [9]. Similar to other silicates, sepiolite exhibits limited dispersion in the polymer matrices due to its hydrophilic nature [10]. In order to overcome this limitation, surface modification with organic molecules such as surfactants, polymers and silane coupling agents has been reported [11-13]. The surface of the sepiolite clay contains large number of silanol (Si-OH) groups, which readily react with organic coupling agents [14]. In a recent study, Chen *et al.* [15] developed sepiolite-polyurethane nanocomposites and reported the tensile strength and elongation at break of the nanocomposites to increase with increasing amount of nanoclay. In a similar study, Bilotti *et al.* [16] observed the sepiolite clay to exhibit higher reinforcing efficiency in polyamide composites compared to MMT clay.

Graphene being an allotrope of carbon has drawn significant research attention as filler for polymer nanocomposites because of its attributes like high surface area, large charge carrier mobility, chemical stability, excellent mechanical properties, as well as high thermal

S. Singh, M. R. Vengatesan^a, Sevim Isci^b and Vikas Mittal^{a,}, ^aThe Petroleum Institute (part of Khalifa University of Science and Technology), Abu Dhabi, UAE; ^bIstanbul Technical University, Istanbul, Turkey*
*Current address: Bletchington, Wellington County, Australia
© 2019 Central West Publishing, Australia

and electrical conductivity [17-20]. Graphene has also been reported to exhibit better properties than other carbon fillers in polymer nano-composites [21-23]. However, the tendency of graphene platelets to aggregate significantly limits the efficient dispersion in polymer ma-trices [24].

More recently, research has focused on exploring the synergistic effects of mixed filler systems on the properties of the polymer ma-trices [25-28]. Graphene has been widely used as one of the compo-nents for the preparation of mixed nanofiller systems by physically mixing with a variety of one dimensional (1D) nanomaterials [29-32]. Such mixing reduces the sheet staking of graphene and generates a synergistic effect on the resultant nanocomposite properties [29]. Noh *et al.* [30] generated cyclic butylene terephthalate nanocompo-sites using pitch based carbon fibers and graphene sheets through physical mixing method and observed a synergistic effect of the fillers on the isotropic thermal conductivity with 82% improved efficiency as compared to the individual fillers. Cho *et al.* [31] developed poly-propylene/polyaniline/graphene nanocomposite by a facile physical mixing of polyaniline fibers and graphene sheets with polypropylene in organic solvent and observed that 10 vol% of co-fillers in the PP matrix resulted in a high dielectric constant of $\varepsilon' \approx 51.8$ with small dielectric loss $\varepsilon'' \approx 9.3 \times 10^{-3}$. Luan *et al.* [32] developed conductive epoxy composites using graphene and silver nanowires *via* simple physical blending method and the nanocomposites exhibited a signif-icant enhancement in the electrical conductivity due to the decreased tunneling resistance (Rt) between the silver nanowires in the pres-ence of the conductive 2-D chemically reduced graphene. Less num-ber of studies have been reported for mixed fibrous clay and gra-phene platelets for the generation of polymer nanocomposites. Kwon *et al.* [25] compared the properties of polyimide nanocomposites generated with organoclay, functionalized graphene and or-ganoclay/functionalized graphene complex. The authors concluded that the clay/graphene complex composite exhibited superior prop-erties compared to other nanocomposites due to the synergistic ef-fect of the binary nanofiller system. Due to potential of achieving en-hanced composite properties using such mixed filler systems, further work involving such systems is required, especially using polyolefins like polypropylene.

In this study, polypropylene nanocomposites were generated us-ing binary filler system containing graphene and surface modified se-piolite. The influence of the mixed filler system on the mechanical, thermal and morphological properties of the composites was studied. Amine functionalized polypropylene (PP-*g*-DA8) was synthesized and used as compatibilizer to obtain chemical interactions with the filler surface.

2.2 Experimental

2.2.1 Materials

Polypropylene (PP) with a trade name of HD51CF was received from Abu Dhabi Polymers (Borouge), UAE. Sepiolite (S) clay was received from Sigma Aldrich. It has a characteristic diffraction peak (d_{001}) at 12.12 Å. The main characteristic bands of sepiolite were reported at 3684 cm^{-1}, 3550 cm^{-1}, 1693 cm^{-1}, 1658 cm^{-1}, 1204 cm^{-1}, 1022 cm^{-1}, 780 cm^{-1}, 691 cm^{-1} and 648 cm^{-1}, which are ascribed to bending of Mg-OH, -OH stretching, OH bending, Si-O-Si bands, bending of Mg-Fe-OH and Mg-OH bending. 1,8-diaminooctane, polypropylene-graft-maleic anhydride (PP-*g*-MA) (M_w ~9,100, M_n ~3,900 and maleic anhydride grafting percentage of 8-10%), acetone, methanol, *o*-xylene, sodium bicarbonate (NaHCO$_3$) and polyurethane (PU) liquid were received from Sigma Aldrich and were used without further purification. Graphene (G) with a trade name of N002-PDR was purchased from Angstron Materials, USA (specific surface area of 400-800 m^2/g, average plane dimension = 1.0-1.2 nm). Ethylene vinyl acetate-*graft*-maleic anhydride (EVA-*g*-MA) copolymer (density 0.962 g/cm^3 and melt flow rate (190 °C/2.16 kg) 1.4 g/10 min) with a trade name of Fusabond C190 was obtained from DuPont.

2.2.2 Surface Modification of Sepiolite

The surface of the sepiolite fibers was modified with PU and EVA-*g*-MA copolymer. For modification with PU, PU was dispersed in deionized (DI) water at a concentration of 1 mg/ml at 70 °C. Sepiolite (2% w/w) was added to the polymer dispersion and shaken for 2 h and ultra-sonicated for 5 min. The content was further stirred overnight. The sample was washed with DI water followed by acetone. The sample was finally dried at 50 °C. The final product was termed as PU-S clay.

Same procedure was used to prepare the EVA-*g*-MA modified sepiolite filler (denoted as EVA*g*MA-S) in toluene solvent. The modified was dried at 100 °C overnight under vacuum to remove the solvent.

2.2.3 Preparation of Amine Functionalized Polypropylene (PP-*g*-DA8) Compatibilizer

The compatibilizer was prepared as per the reported literature [33]. In brief, 10 g of PP-g-MA and 40 g of 1, 8-diaminooctane were added together in N_2 atmosphere. The mixture was heated to 130 °C and stirred for 30 min. Subsequently, it was allowed to cool at room temperature. The content was washed with 5 wt% $NaHCO_3$ solution followed by water and acetone. The material was dried under vacuum for 12 h at 50 °C.

2.2.4 Generation of PP Nanocomposites

The nanocomposites were prepared using solution blending of PP and PP-*g*-DA8 with the fillers. Typically, 400 mL of xylene was placed in a round bottom flask and 0.5 g of EVA*g*MA-S was added to it, followed by 1 h ultra-sonication. 1.0 g of PP-*g*-DA8 and 8.5 g of PP were subsequently added to the flask and stirred at 130 °C for 2 h with a reflux condenser. The solvent was then evaporated at 100 °C under vacuum (Scheme 2.1a). The prepared composite, named as P/5%EVA*g*MA-S, was molded using injection molding (Thermo Scientific) at 190 °C with 400 bar for 10 s to obtain test specimens. The remaining nanocomposites were similarly prepared using 10 wt% EVA*g*MA-S, 5 wt% PU-S, 10 wt% PU-S, 2.5 wt% EVA*g*MA-S + 2.5 wt% G, 2.5 wt% PU-S + 2.5 wt% G, 5 wt% EVA*g*MA-S + 5 wt% G and 5 wt% PU-S + 5 wt% G, with PP-*g*-DA8 concentration fixed at 10 wt%. The nanocomposites were respectively named as P/10%EVA*g*MA-S, P/5%PU-S, P/10%PU-S, P/2.5%EVA*g*MA-S/2.5%G, P/2.5%PU-S/2.5%G, P/5%EVA*g*MA-S/5%G and P/5%PU-S/5%G (Scheme 2.1b). In order to confirm the covalent interaction between the nanofiller and compatibilizer, hybrid of PP-*g*-DA8 and modified sepiolite was also generated. For this, 0.5 g of modified sepiolite was dispersed in 50 mL of dimethyl formamide using an ultra-sonication bath at 30 °C for 1 h. Subsequently, 0.5 g of PP-*g*-DA8 was added to the dispersion and the mixture was stirred for 2 h at 110 °C. The hybrid was obtained by evaporation of the solvent under vacuum at 100 °C.

2.2.5 Characterization

IR spectra of the samples were collected on Nicolet iS10 spectrometer equipped with SmartiTR diamond ATR accessory (angle of incidence

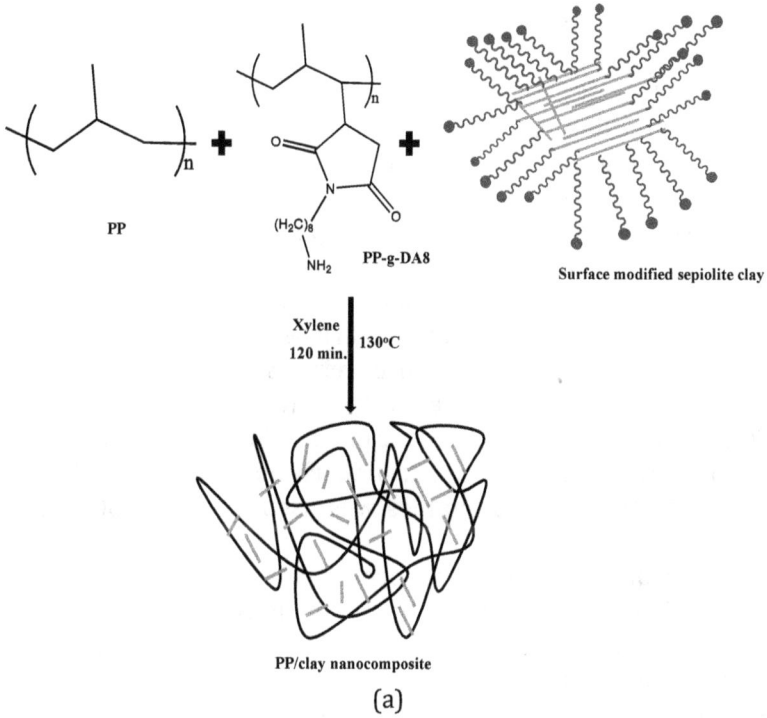

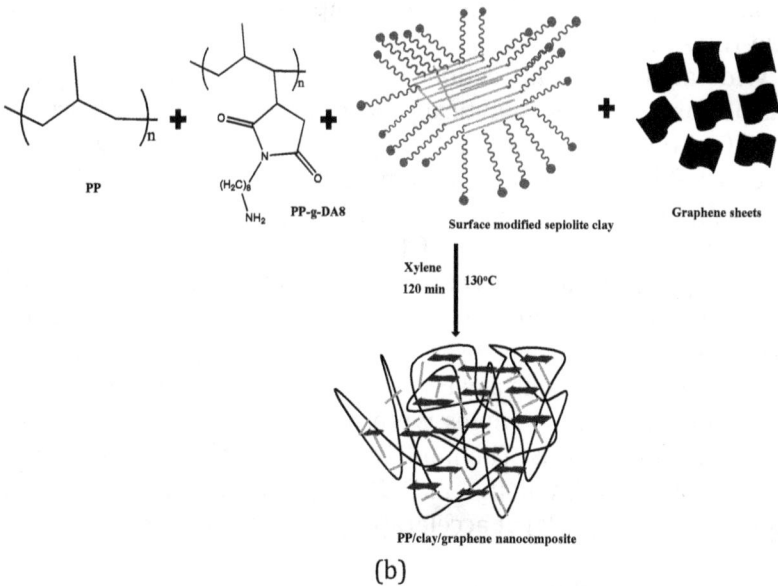

Scheme 2.1 Schematic of the generation of (a) polypropylene/clay and (b) polypropylene/clay/graphene nanocomposites.

of 45°), DTGS KBr detector and KBr beam splitter. It had a diamond ATR crystal (index of refraction = 2.4 at 1000 cm^{-1}) and a depth permeation of 2 µm at 1000 cm^{-1} for sample with refractive index of 1.5. The spectra of the samples were recorded using OMNIC software in the spectral range of 4000-600 cm^{-1} with a resolution of 4 cm^{-1} from 32 scans.

Thermal properties of the nanocomposites were recorded using Discovery Thermogravimetric Analyzer (TGA) in nitrogen medium. A temperature range from 35 to 700 °C at a heating rate of 10 °C/min was used. Differential scanning calorimetric thermograms of the composites were recorded on a Discovery DSC under nitrogen atmosphere. The scans were obtained from 35 to 200 °C and from 200 to 35 °C using heating and cooling rate of 10 °C/min. The second heating and cooling runs were carried out similarly and used to calculate melt enthalpy. 5-10 mg of the samples was used for both DSC and TGA analysis.

Mechanical testing of the nanocomposites was carried out using Instron universal testing machine following ASTM D 638 standard. The injection molded dumbbell-shaped samples were used, and an average of five specimens was recorded. A loading rate of 10 mm/min was used, and the tests were carried out at room temperature. The tensile strength and modulus of the composites were calculated using Blue-Hill Analysis software.

Wide-angle X-ray diffraction (WAXRD) analysis of the composites was performed using analytical powder diffractometer (X'Pert PRO) using Cu Kα radiation (λ = 1.5406 Å) in reflection mode. A zero-background holder was used to minimize the noise. The samples were step-scanned from 2θ = 5-60° at room temperature using a step size of 2θ = 0.02° and a step time of 10 s. For the transmission electron microscopy (TEM) analysis of the ultra-thin film samples, Philips CM 20 (Philips/FEI, Eindhoven) electron microscope at 120 and 200 kV accelerating voltages was used. The samples were prepared by microtoming the film using a glass cutter followed by a diamond cutter. 70-90 nm thick film samples were micro toned and supported on TEM grids for analysis. The modified sepiolite fillers were analyzed in Hitachi field emission in-lens S-900 high resolution scanning electron microscope (SEM) at accelerating voltages of 10-20 kV.

2.3 Results and Discussion

Figure 2.1 demonstrates the SEM images of the pristine and modified

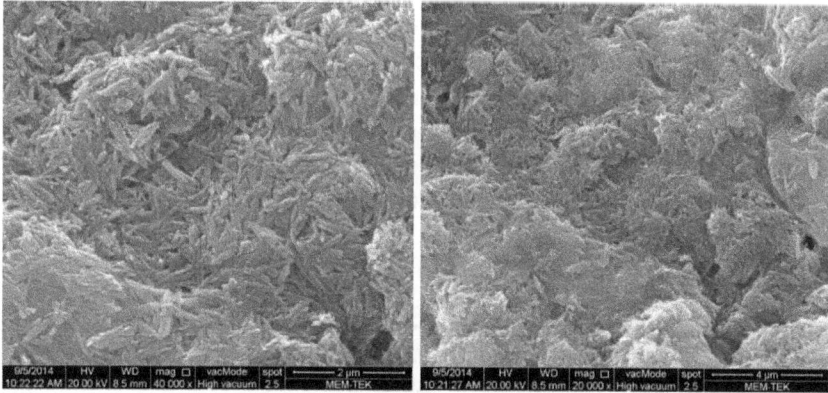

(a)

(b)

(c)

Figure 2.1 SEM images of (a) pristine sepiolite, (b) PU-S and (c) S-EVAgMA-S at two different magnifications.

sepiolite. The pristine particles were observed to be uniformly dispersed as primary particles without forming any large sized aggregates. On the other hand, the filler particles were observed to form large aggregates in these modified sepiolite materials. However, it should be noted that the observed aggregates may be only loosely held aggregated structures, which may be easily broken when blended with polymers.

Figure 2.2a shows the FT-IR spectra of PP-g-DA8, EVAgMA-S and

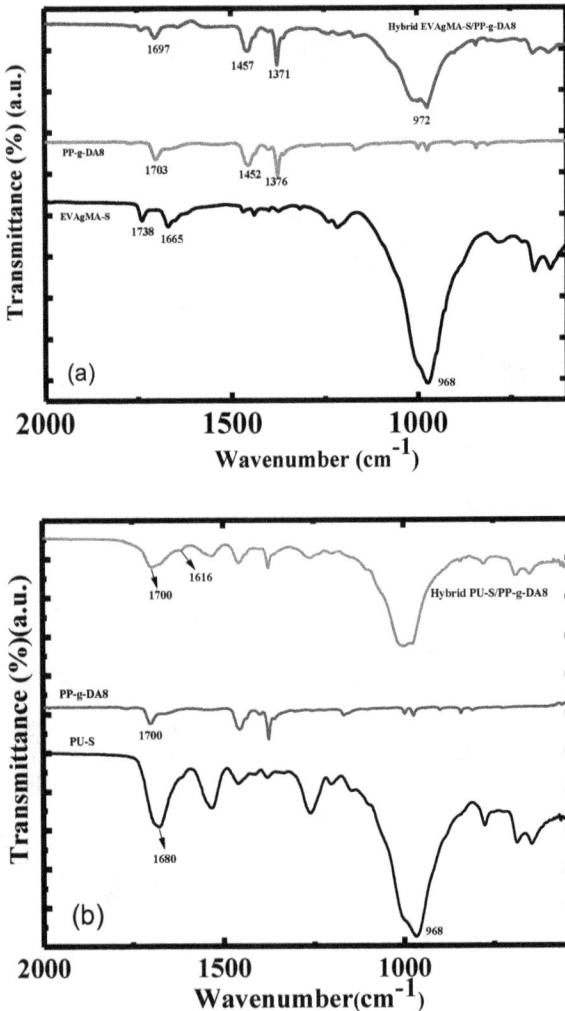

Figure 2.2 FTIR spectra of a) EVAgMA-S, PP-g-DA8 and hybrid EVAgMA-S/PP-g-DA8, b) PU-S, PP-g-DA8 and hybrid PU-S/PP-g-DA8.

PP-*g*-DA8/EVA*g*MA-S hybrid. The spectrum of EVA*g*MA-S exhibited maleic anhydride group with characteristic absorption bands at 1738 and 1665 cm^{-1} associated with the carbonyl C=O symmetric and asymmetric stretching bands. PP-*g*-DA8 exhibited an absorption band at 1703 cm^{-1} attributed to primary amide C=O stretching and another absorption band at 1457 cm^{-1} associated with N–H secondary amine stretching. In the hybrid of EVA-*g*-MA and PP-*g*-DA8, most of the carbonyl C=O symmetric and asymmetric stretching bands at 1738 cm^{-1} vanished indicating the reaction between the EVA*g*MA-S and the compatibilizer [33,34]. In addition, a broad intense peak appeared at 972 cm^{-1} representing the magnesium silicate of sepiolite. Figure 2.2b represents the FT-IR spectra of PP-*g*-DA8, PU-S and PP-*g*-DA8/PU-S hybrid. The hybrid of PU-S with PP-*g*-DA8 did not exhibit any chemical interaction and only a small peak was observed at 1616 cm^{-1} due to the weak hydrogen bonding between the C=O of PU-S and N-H of PP-*g*-DA8 [35], thus, representing a physical blend. Figure 2.3a shows the FT-IR spectra of the 10% PU-S and EVA*g*MA-S filled PP nanocomposites. Same reaction between EVA*g*MA-S and PP-g-DA8 was observed to occur when PP was added. In addition, the incorporation of graphene resulted a characteristic hump like peak at 1583 cm^{-1}, which corresponded to the C=C stretching frequency of graphene (Figure 2.3b) [36].

Table 2.1 presents the melting and crystallization properties of pure PP, PP/PP-*g*-DA8 blend and nanocomposites. The degree of crystallinity (X_c) was calculated using the melting enthalpy (ΔH_m) of the 100 % crystalline PP (209 J/m) [37]. The incorporation of 10 wt% PP-*g*-DA8 reduced the melt enthalpy and the degree of crystallinity of the polymer. These results suggested that the polar amine-PP restricted the ability of the pure PP chains to form crystals and the amine-PP acted as a plasticizer in the blend [33]. The crystallinity of nanocomposites increased with increasing percentage of sepiolite. Both surface modifications for sepiolite resulted in a significant increase in the % crystallinity of the polymer in the composites, thus, confirming the nucleation effect. In the case of hybrid filler composites, addition of clay and graphene in the PP matrix resulted in slightly enhanced % crystallinity as compared to clay composites due to enhanced nucleating effect in the presence of graphene.

The thermal degradation temperature for 10 wt% and 40 wt% loss as well as char yield value at 700 °C are reported in Table 2.2. It was observed that the addition of compatibilizer resulted in an increase in the degradation temperature of the PP matrix. For example,

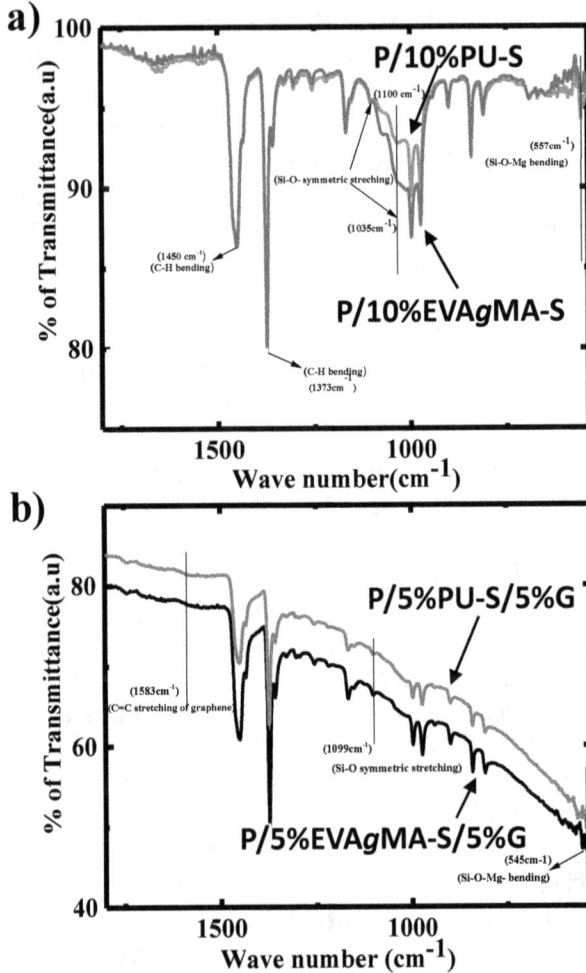

Figure 2.3 FTIR spectra of a) PP/clay nanocomposites, b) PP/clay/graphene nanocomposites.

the 10 wt% loss temperature for pure PP was observed to be 374 °C in nitrogen atmosphere, which was 43 °C less than the PP/PP-*g*-DA8 matrix. As the compatibilizer had both imide linkage and terminal amine group, it was expected to impart enhanced thermal stability. In addition, clay and hybrid clay-graphene filled nanocomposites exhibited almost similar 10% and 40% weight loss temperatures. The char yield of the nanocomposites increased with increase in the nanoclay fraction as compared to pure PP and PP/PP-*g*-DA8 blend, which can be attributed to the role of nanoclay as thermal shield in the polymer nanocomposites, thus, increasing the char yield. The hybrid clay/graphene nanofillers did not exhibit any significant synergistic effect on the char yield of the PP/clay/graphene nanocomposites.

Table 2.1 Calorimetric properties of nanocomposites

Sample	T_c (°C)	T_m (°C)	ΔH_m (J/g)	ΔH_c (J/g)	Xc (%)
PP	128	167	17	24	35
PP/PP-*g*-DA8	128	168	11	23	26
P/5%EVAgMA-S	128	167	17	22	41
P/5%PU-S	128	166	18	22	42
P/10%EVAgMA-S	129	167	17	21	43
P/10%PU-S	127	167	18	20	45
P/2.5%EVAgMA-S/2.5%G	127	169	19	20	45
P/2.5%PU-S/2.5%G	130	167	20	21	47
P/5%EVAgMA-S/5%G	130	168	17	19	43
P/5%PU-S/5%G	130	169	19	18	47

Table 2.2 Thermal stability and char yield of PP/clay and PP/clay/graphene nanocomposites

Sample	T_d^{10} (°C)	T_d^{40} (°C)	Char yield @ 700 °C
PP	386	414	1.3
PP/PP-*g*-DA8	429	456	1.4
P/5%EVAgMA-S	431	457	3.3
P/5%PU-S	432	458	2.2
P/10%EVAgMA-S	433	458	7.0
P/10%PU-S	431	460	5.0
P/2.5%EVAgMA-S/2.5%G	429	458	3.4
P/2.5%PU-S/2.5%G	429	458	3.2
P/5%EVAgMA-S/5%G	439	460	9.2
P/5%PU-S/5%G	440	462	8.0

The mechanical properties of nanocomposites are summarized in Table 2.3. Incorporation of 10 wt% compatibilizer to the PP matrix resulted in a significant decrease in the tensile modulus and strength due to the plasticization effect. Subsequently, the addition of EVA*g*MA-S to the polymer resulted in an increasing trend for both modulus and strength, with increase in clay content. For instance, at 10% EVA*g*MA-S content, the modulus and strength were enhanced

Table 2.3 Mechanical properties of PP/clay and PP/clay/graphene hybrid nanocomposites

Composites	Max. load	UTS [MPa]	Modulus [MPa]	Strain at break (%)
PP	212± 2	36 ± 2	891 ± 76	13 ± 2
PP/PP-*g*-DA8	194 ± 16	32 ± 3	784 ± 11	8 ± 3
P/5%EVA*g*MA-S	201 ± 4	33 ± 2	852 ± 54	6 ± 2
P/5%PU-S	152 ± 27	25 ± 4	838 ± 34	4 ± 1
P/10%EVA*g*MA-S	215 ± 3	36 ± 2	917 ± 32	6 ± 2
P/10%PU-S	190 ± 7	32 ± 2	911 ± 37	5± 1
P/2.5%EVA*g*MA-S/2.5%G	325 ± 12	41± 2	944 ± 54	7± 1
P/2.5%PU-S/2.5%G	322 ± 17	41 ± 2	935 ± 49	8 ± 2
P/5%EVA*g*MA-S/5%G	328 ± 5	43 ± 2	872 ± 41	8 ± 2
P/5%PU-S/5%G	290 ± 11	37 ± 2	867 ± 65	5 ± 1

by 16% and 11% respectively as compared to the PP/PP-*g*-DA8 blend. The reaction of the maleic anhydride functional group on the filler surface with the amine functionalized compatibilizer, thus, forming strong imide linkage between the compatibilizer and the filler phase, would have contributed to the enhanced mechanical performance of the composites. This was further confirmed from the mechanical performance of the nanocomposites with PU-S filler. The nanocomposites with PU-S exhibited significant decrease in the tensile strength, and the modulus was enhanced to a lesser extent as compared to EVA*g*MA-S. In comparison with the chemical linkage of EVA*g*MA-S with the compatibilizer, intermolecular H-bonding between the PU-S and PP-*g*-DA8 along with the plasticization effect of the urethane linkage in PU would have resulted in such performance. In the case of hybrid filler nanocomposites, a remarkable synergistic effect on both modulus and strength was observed for P/2.5%EVA*g*MA-S/2.5%G and P/2.5%PU-S/2.5%G. However, P/5%EVA*g*MA-S/5%G and P/5%PU-S/5%G composites exhibited a decrease in the tensile modulus probably due to filler aggregation. The tensile strength for P/5%PU-S/5%G was also observed to decrease, though the values were still higher than the corresponding sepiolite composites. It is well known that the nanofillers have a threshold limit of the loading amount in the polymer matrix. It is probable that the sepiolite phase failed to reduce the stacking of graphene sheets at higher concentration, as the filler materials have a tendency to aggregate due to van der Waals forces at higher contents. Thus, for the mixed filler system, the higher filler loading induced the aggregation due to weaker interfacial interaction between the fillers.

XRD spectra of pure PP, PP/PP-*g*-DA8 blend and nanocomposites (Figure 2.4) contained six major reflections of α-phase at 14.6°, 17.7°,

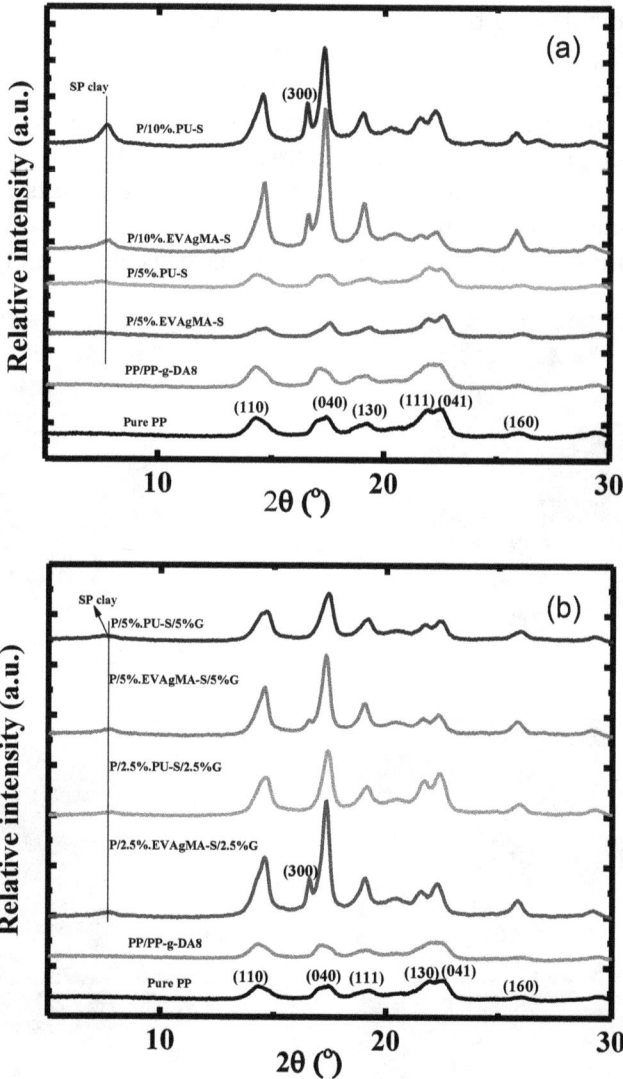

Figure 2.4 XRD spectra of a) PP/clay nanocomposites, b) PP/clay/graphene nanocomposites.

19.0°, 21.9°, 22.6° and 26.0° corresponding to planes (110), (040), (130), (111), (041) and (060) respectively. The (110) α and (040) α peaks were compared taking (110) α as a reference peak. The increase in the ratio of intensities between the two peaks confirmed an increase in the degree of crystallinity, which also supported the DSC results. The addition of 10 wt% sepiolite induced the β-crystallinity of PP at the plane (300). The β-crystalline phase is thermally less stable as compared to the α-phase of PP under normal crystallization conditions [39]. The presence of nucleating agents can also assist the formation of β-form in PP. However, in the case of sepiolite/graphene hybrid composites, the β- crystal formation was only observed in the EVA*g*MA-S/graphene nanocomposites. As the β-crystal formation improves the mechanical properties of PP, these results coincided with the mechanical properties of PP nanocomposites.

TEM analysis of the nanocomposites is presented in Figure 2.5.

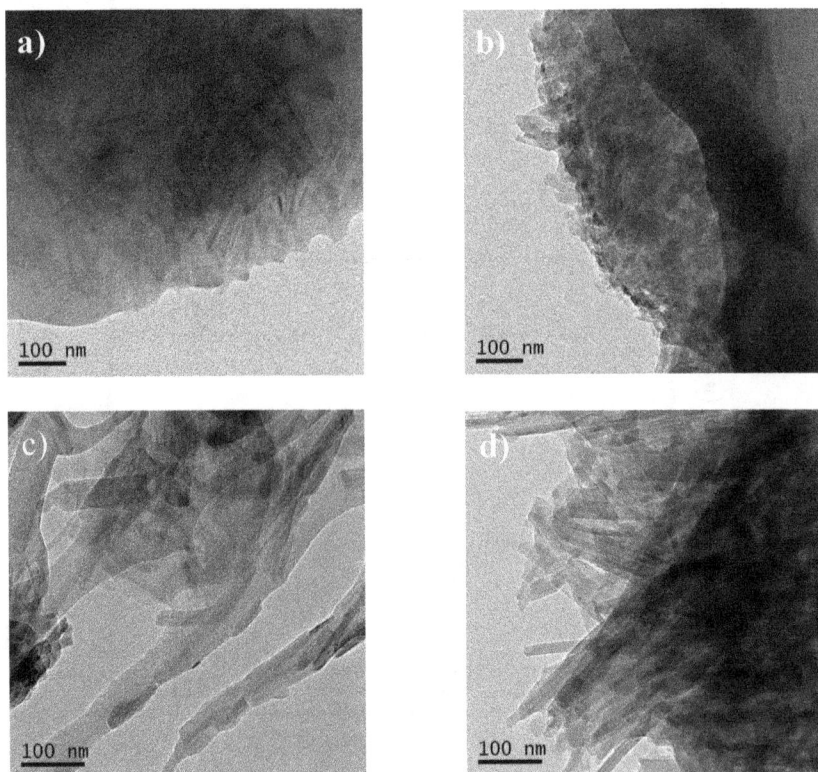

Figure 2.5 TEM images of (a) P/5%PU-S, (b) P/5%EVAgMA-S, (c) P/2.5%PU-S/2.5%G and (d) P/5%EVAgMA-S/5%G nanocomposites.

Figure 2.5a, corresponding to P/5%PU-S nanocomposite, exhibited uniform distribution of filler in the PP matrix, however, intercalated clay layers was also observed. Thus, though the compatibilizer led to secondary H-bonding with sepiolite and resulted in overcoming the energy barrier to form randomized delamination of sepiolite, complete delamination was, however, not achieved. Figure 2.5b shows the TEM image of P/5%EVA*g*MA-S nanocomposite. The sepiolite phase was observed to be uniformly distributed in the PP matrix and exhibited an exfoliated morphology. Chemical interaction of the compatibilizer with the filler surface resulted in an enhanced degree of filler delamination, which also translated into mechanical performance of the composites. Figure 2.5c and 2.5d show the images for the nanocomposites with 2.5wt%EVA*g*MA-S+2.5%G and 5%EVA*g*MA-S+5%G respectively. No filler aggregation was observed, and the particles were distributed at the nanoscale in the polymer matrix. Some degree of filler aggregation was also observed in the composites with higher filler content.

2.4 Conclusions

In the present study, PP nanocomposites were developed using organo-modified sepiolite clay (EVA*g*MA-S and PU-S) as well as graphene-modified sepiolite clay hybrid in the presence of amine functionalized PP as a compatibilizer.

EVA*g*MA modified sepiolite was observed to chemically interact with the compatibilizer. TEM images displayed the uniform dispersion of nanofillers, especially EVA*g*MA-S, in the PP nanocomposites at lower filler content. An improved thermal stability was observed in the composites along with an increase in the char yield. DSC studies showed that the nanofillers acted as weak nucleating agents, thus, resulting in an increase in the crystallinity of the polymer matrix. Further from XRD, an increase in the degree of crystallinity was confirmed when the ratio of (110) α and (040) α peaks was compared, with (110) α as the reference peak. An improvement in Young's modulus was observed for the nanocomposites, especially with EVA*g*MA-S clay containing composites, due to the covalent interaction of the filler surface with the compatibilizer, thus, resulting in better filler dispersion.

Addition of graphene further enhanced the tensile modulus, thus, confirming the synergistic effect of the mixed filler system on the polymer properties.

Acknowledgment

The authors sincerely thank the PI Research Centre, The Petroleum Institute for the financial support.

References

1. Kawasumi, M., Hasegawa, N., Kato, M., Usuki, A., and Okada, A. (1997) Preparation and mechanical properties of polypropylene-clay hybrids. *Macromolecules,* **30**, 6333-6338.
2. Garcia-Lopez, D., Picazo, O., Merino, J. C., and Pastor, J. M. (2003) Polypropylene-clay nanocomposites: effect of compatibilizing agents on clay dispersion. *European Polymer Journal,* **39**, 945-950.
3. Liu, X., and Wu, Q. (2001) PP/clay nanocomposites prepared by grafting melt-intercalation. *Polymer,* **42**, 10013-10019.
4. Vengatesan M. R., and Mittal, V. (2015) Surface modification of nanomaterials for application in polymer nanocomposite: An overview. In: *Surface Modification of Nanoparticle and Natural Fiber Fillers*, Wiley-VCH, Germany, pp. 1-28.
5. Saravanan, S., Gowda, K. M. A., Varman, K. A., Ramamurthy, P. C., and Madras, G. (2015) In-situ synthesized poly (vinyl butyral)/MMT-clay nanocomposites: The role of degree of acetalization and clay content on thermal, mechanical and permeability properties of PVB matrix. *Composites Science and Technology,* **117**, 417-427.
6. Sapalidis, A. A., Katsaros, F. K., Steriotis, T. A., and Kanellopoulos, N. K. (2012) Properties of poly (vinyl alcohol)- Bentonite clay nanocomposite films in relation to polymer-clay interactions. *Journal of Applied Polymer Science,* **123**, 1812-1821.
7. Zheng, S., Wang, T., Liu, D., Liu, X., Wang, C., and Tong, Z. (2013) Fast deswelling and highly extensible poly(N-isopropylacrylamide)-hectorite clay nanocomposite cryogels prepared by freezing polymerization. *Polymer,* **54**, 1846-1852.
8. Mittal, V., and Al Zaabi, K. (2013) Biodegradable polyester nanocomposites: Phase miscibility and properties. *Journal of Applied Polymer Science,* **130**(1), 516-525.
9. Garcia-Lopez, D., Fernandez, J. F., Merino, J. C., Santaren, J., and pastor, J. M. (2010) Effect of organic modification of sepiolite for PA6 polymer/organoclay nanocomposites. *Composites Science and Technology,* **70**, 1429-1436.
10. Bergaya, F., Then, B. K. G., and Lagaly, G. (2006) *Handbook of Clay Science*, Elsevier, Netherlands.
11. Samakande, A., Hartmann, P. C., Cloete, V., and Sanderson, R. D. (2007) Use of acrylic based surfmers for the preparation of exfoliated polystyrene clay nanocomposites. *Polymer,* **48**, 1490-1499.

12. Mansoori, Y., Atghia, S. V., Zamanloo, M. R., Imanzadeh, Gh., and Si-rousazar, M. (2010) Polymer-clay nanocomposites: Free-radical grafting of polyacrylamide onto organophilic montmorillonite. *European Polymer Journal*, **46**, 1844-1853.
13. Choi, Y. Y., Lee, S. H., and Ryu, S. H. (2009) Effect of silane function-alization of montmorillonite on epoxy/montmorillonite nanocom-posite. *Polymer Bulletin*, **63**, 47-55.
14. Grim, R. E. (1962) *Applied Clay Mineralogy*, McGraw-Hill, USA.
15. Chen, H., Zheng, M., Sun, H., and Jia, Q. (2007) Characterization and properties of sepiolite/polyurethane nanocomposites. *Materials Science and Engineering A*, **445-446**, 725-730.
16. Bilotti, E., Zhang, R., Denge, H., Quero, F., Fischer, H. R., and Peijs, T. (2009) Sepiolite needle like clay for PA6 nanocomosites: An alter-native to layered silicates?. *Composites Science and Technology*, **69**, 2587-2595.
17. Novoselov, K. S., Geim, A. K., Morozov, S. V., Jiang, D., Katsnelson, M. I., Grigorieva, I. V., Dubon, S. V., and Firsov, A. A. (2005) Two-dimen-sional gas of massless Dirac fermions in graphene. *Nature*, **438**, 197-200.
18. Novoselov, K. S., Jiang, Z., Zhang, Y., Morozov, S. V., Stormer, H. L., Zeitler, U., Mann, J. C., Boebinger, G. S., Kim, P., and Geim, A. K. (2007) Room-temperature quantum hall effect in graphene. *Science*, **315**, 1379.
19. Wang, G., Yang, J., Park, J., Gou, X., Wang, B., and Liu, H. (2008) Facile synthesis and characterization of graphene nanosheets. *Journal of Physical Chemistry C*, **112**, 8192-8195.
20. Wang, G., Shen, X., Wang, B., Yao, J., and Park, J. (2009) Synthesis and characterisation of hydrophilic and organophilic graphene nanosheets. *Carbon*, **47**, 1359-1364.
21. Li, X., Wang, X., Zhang, L., Lee, S., and Dai, H. (2008) Chemically de-rived, ultrasmooth graphene nanoribbon semiconductors. *Science*, **319**, 1229-1231.
22. Eda, G., and Chhowalla, M. (2009) Graphene-based Composite Thin Films for Electronics. *Nano Letters*, **9**, 814-818.
23. Liang, J., Xu, Y., Huang, Y., Zhang, L., Wang, Y., and Ma, Y. (2009) In-frared-Triggered Actuators from Graphene-Based Nanocomposites. *Journal of Physical Chemistry C*, **113**, 9921-9927.
24. Kim, H., and Macosko, C. W. (2009) Processing-property relation-ships of polycarbonate/graphene composites. *Polymer*, **50**, 3797-3809.
25. Kwon, K., and Chang, J.-H. (2015) Comparison of the properties of polyimide nanocomposites containing three different nanofillers: organoclay, functionalized graphene, and organoclay/ functional-ized graphene complex. *Journal of Composite Materials*, **49**, 3031-3044.

26. Paszkiewicz, S., Kwiatkowska, M., Rosłaniec, Z., Szymczyk, A., Jotko, M., and Lisiecki, S. (2015) The influence of different shaped nano-fillers (1D, 2D) on barrier and mechanical properties of polymer hybrid nanocomposites based on PET prepared by in situ polymerization. *Polymer Composites,* **37**(7), 1949-1959.

27. Pradhan, B., Roy, S., Srivastava, S. K., and Saxena, A. (2015) Synergistic effect of carbon nanotubes and clay platelets in reinforcing properties of silicone rubber nanocomposites. *Journal of Applied Polymer Science,* **132**, 41818.

28. Kim, H., Abdala, A. A., and Macosko, C. W. (2010) Graphene/Polymer Nanocomposites. *Macromolecules,* **43**, 6515-6530.

29. Tang, Z., Wei, Q., Lin, T., Guo, B., and Jia, D. (2013) The use of a hybrid consisting of tubular clay and graphene as a reinforcement for elastomers. *RSC Advances,* **3**, 17057-17064.

30. Noh, Y. J., and Kim, S. Y. (2015) Synergistic improvement of thermal conductivity in polymer composites filled with pitch based carbon fiber and graphene nanoplatelets. *Polymer Testing,* **45**, 132-138.

31. Cho, S. Kim, M. Lee, J. S., and Jang, J. (2 015) Polypropylene/polyaniline nanofiber/reduced graphene oxide nanocomposite with enhanced electrical, dielectric, and ferroelectric properties for a high energy density capacitor. *ACS Applied Materials & Interfaces,* **7**, 22301-22314.

32. Luan, V. H., Tien, H. N., Cuong, T. V., Kong, B.-S., Chung, J. S., Kim, E. J., and Hur, S. H. (2012) Novel conductive epoxy composites composed of 2-D chemically reduced graphene and 1-D silver nanowire hybrid fillers. *Journal of Materials Chemistry,* **22**, 8649-8653.

33. Sanchez-Valdes, S., Mendez-Nonell, J., Medellin-Rodriguez, F. J., Ramirez- Vargas, E., Martinez-Colunga, J. G., Ramos de Valle, L.F., Mondragon-Chaparro, M., Lopez-Quintanilla, M. L., and Garcia-Salazar, M. L. (2010) Evaluation of different amine- functionalized polyethylene as compatibilizer for polyethylene film nanocomposites. *Polymer International,* **59**, 704-711.

34. Cui, L., and Paul, D. R. (2007), Evaluation of amine functionalized polypropylene as compatibilizers for polypropylene nanocomposites, *Polymer,* **48**, 1632-1640.

35. Deka, H., and Karak, N. (2009) Vegetable oil-based hyperbranched thermosetting polyurethane/clay nanocomposites. *Nanoscale Research Letters,* **4**, 758-765.

36. Vengatesan, M. R., Shen, T. Z., Alagar, M., and Song, J.-K. (2016) A facile chemical reduction of graphene-oxide using p-toluene sulfonic acid and fabrication of reduced graphene-oxide film. *Journal of Nanoscience and Nanotechnology,* **16**, 327-332.

37. Pedrazzoli, D., Khumalo, V. M., Kocsis, J. K., and Pegoretti, A. (2014) Thermal, viscoelastic and mechanical behavior of polypropylene with synthetic boehmite alumina nanoparticles. *Polymer Testing,*

35, 92-100.

38. Zaman, I., Manshoor, B., Khalid, A., Meng, Q., and Araby, S. (2014) Interface modification of clay and graphene platelets reinforced epoxy nanocomposites: A comparative study. *Journal of Materials Science*, **49**, 5856-5865.

39. Bao, S. P., and Tjong, S. C. (2 007) Impact essential work of fracture of polypropylene/montmorillonite nanocomposites toughened with SEBS-g-MA elastomer. *Composites Part A*, **38**, 378-38727.

3

Polypropylene/Cyclohexylethylene-*co*-Ethylene Blends and Polypropylene/Cyclohexylethylene-*co*-Ethylene/ Graphene Nanocomposites

3.1 Introduction

Polymer blending and composite generation are two of the widely acceptable techniques used to produce high performance polymeric materials. In comparison to the synthesis of new monomers or polymers, polymer blending is beneficial to upgrade the performance and processability of polymers in a simple, effective and economic way. Also, it presents benefits in terms of polymer recycling and waste management [1-4]. Polymer composites are engineering hybrid materials based on filler reinforced polymer matrices and have displayed significant potential to replace conventional materials in many applications areas owing to weight saving, strength, dimensional, thermal and chemical stability, flame retardancy, ease of processing and handling, cost effectiveness, etc. [5,6]. Polymer nanocomposites are an advanced class of promising composites, with one dimension of the filler phase lower than 100 nm. The nanomaterials like graphene, carbon nanotubes, nanoclay, etc., improve the properties of polymers significantly even at very low loadings in comparison to other conventional polymer composites [7-10]. The dispersion state of these nanomaterials in the polymer matrices is the critical factor which decides the extent of interfacial interactions between the components and, thus, final properties [11].

On account of its superior mechanical, thermal and barrier properties, along with unique structural features, graphene has received extensive research attention for designing high performance polymer nanocomposites. Graphene is a two dimensional nanomaterial with a hexagonally organized sp^2 bonded sheet morphology and exhibits structural similarity to silicates as well as chemical similarity to carbon nanotubes [12-15]. Typically, graphene is successfully sourced

Anish M. Varghese and Vikas Mittal, The Petroleum Institute (part of Khalifa University of Science and Technology), Abu Dhabi, UAE*
*Current address: Bletchington, Wellington County, Australia
© 2019 Central West Publishing, Australia

from graphite by means of four different routes, namely thermal reduction, chemical reduction, chemical vapor decomposition and micro-mechanical expansion [16-18]. Achievement of effective graphene dispersion in the polymer matrices, especially in the case of non-polar polymers, is a challenging task [19,20]. It can be overcome by reducing the polarity difference between the polymer and filler phases. In this context, the use of compatibilizers possessing both hydrophilic and hydrophobic components is one of the effective ways to improve the compatibility of graphene and its derivatives with polymer matrices in a non-covalent manner [21-23].

Polypropylene (PP) is one of the most commonly used commodity thermoplastic, which has attracted significant research interest because of its versatile properties, low cost and easy processability. The unique properties of PP drive its suitability for a wide range of applications such as automotive parts, household implements, building and construction materials, medical devices, electrical and electronic devices, other industrial materials, etc. [24-27]. Considering the rising importance of PP, a number of research studies have reported the incorporation of nanomaterials, especially graphene, to further enhance its properties.

In a recent study, Haghnegahdar *et al.* [28] studied the influence of graphene on the fracture toughness of ethylene propylene diene monomer (EPDM) toughened PP. The fracture toughness of the PP/EPDM blend was observed to improve after graphene addition, and the degree of improvement was higher for few-layer graphene (FLG) platelets in comparison with multi-layer graphene platelets due to the improved dispersion state. Also, the application of dynamic vulcanization was observed to be beneficial to further strengthen the fracture toughness of PP/EPDM/graphene nanocomposites. In another study, the authors reported enhanced electrical and thermal conductivity as well as thermal stability for graphene reinforced PP/EPDM blend nanocomposites [29]. The authors also made use of PP functionalized graphene and observed improved graphene dispersion in the PP/EPDM blend matrix. It helped to strengthen the interfacial bonding, which correspondingly exhibited enhanced tensile strength and modulus as well as reduced damping [30].

Parameswaranpillai *et al.* [31] employed styrene-(ethylene-*co*-butylene)-styrene triblock copolymer (SEBS) to obtain improved interfacial interactions between graphene and PP, which correspondingly resulted in improved toughness and thermal stability, attributed to the diminished graphene flocculation tendency. In another recent study, the authors reported the generation of PP/polysty-rene/SEBS/graphene nanocomposites possessing superior tough-ness, stiffness and thermal stability for effective automotive applica-tions [32]. Li *et al.* [33] also used PP-graft-maleic anhydride (PP-*g*-MA) as a compatibilizer for developing PP/graphene nanocompo-sites. Due to the influence of PP-*g*-MA, the graphene nanosheets were dispersed homogeneously in the PP matrix, which resulted in im-proved static and dynamic mechanical behavior, electrical properties and thermal stability.

In this study, cyclohexylethylene-*co*-ethylene (CHEE) copolymer was used to develop PP blends and FLG reinforced PP blend nano-composites using direct melt blending technique. The effect of vary-ing amounts of CHEE and graphene on the mechanical, thermal and morphological characteristics of PP was evaluated using universal testing (UTM) analysis, Izod impact test, differential scanning calo-rimetry (DSC), thermogravimetric analysis (TGA), laser flash analysis (LFA), wide angle X-ray diffraction (WAXD), Fourier-transform infra-red (FTIR) spectroscopy, Raman spectroscopy and scanning electron microscopy (SEM). The ultimate objective of the study was to explore the benefit of graphene to simultaneously enhance tensile modulus, tensile strength, impact strength, thermal stability and thermal con-ductivity in the presence of CHEE copolymer.

3.2 Experimental

3.2.1 Materials

PP matrix was a homopolymer grade with a melt flow rate of 8 g/10 min at 230 °C/2.16 kg and a density in the range of 900-910 kg.m^{-3} from Abu Dhabi Polymers (Borouge), UAE. Cyclohexylethylene-*co*-ethylene with cyclohexylethylene content of 60%, average M_w of ~90 kDa, polydispersity index (PDI) of 1.12 and glass transition tempera-ture (T_g) of 29 °C was procured from Signa Aldrich. FLG grade N002-PDR with carbon content of 96.3%, average thickness of 1.0-1.2 nm, average length of ≤10 μm and surface area of 400-800 m^2.g^{-1} was pur-chased from Angstron Materials, USA.

3.2.2 Preparation of PP/CHEE Blends

PP/CHEE blends with weight compositions of 95/5, 90/10, 85/15 and 80/20 were melt blended in a mini conical co-rotating twin screw extruder (HAAKE MiniLab, Thermo Scientific). The melt blending was performed at 180 °C for 15 minutes with a screw speed of 100 rpm. For the preparation of dumbbell and rectangular bar shaped specimens for respective tensile and Izod impact analysis, mini lab-scale injection molding machine (HAAKE MiniJet PRO, Thermo Scientific) was used. Cylinder temperature of 180 °C, mold temperature of 125 °C, injection pressure of 430 bar for 10 s and post pressure of 500 bar for 6 s were used.

3.2.3 Preparation of PP/CHEE/Graphene Nanocomposites

PP blend with 10 wt% CHEE (PP/10CHEE) was used as the matrix material for the fabrication of FLG nanocomposites. To study the effect of FLG content, PP/10CHEE blend nanocomposites with 0.5, 1, 3 and 5 wt% FLG were fabricated using mini conical co-rotating twin screw extruder (HAAKE MiniLab, Thermo Scientific) by following the same compounding parameters as used in the case of PP/CHEE blends. The resulting nanocomposites were processed to form dumbbell, rectangular bar and disc shaped specimens using mini lab-scale injection molding machine (HAAKE MiniJet PRO, Thermo Scientific), using the same conditions as mentioned above.

3.2.4 Characterization of Blends and Nanocomposites

Resil impactor (Ceast) equipped with a hammer energy of 4 J and a speed of 3.64 m.s^{-1} was used to analyze the un-notched Izod impact strength of blends and nanocomposites as per ISO 180. The samples were analyzed at room temperature using 80 mm x 10 mm x 4 mm rectangular bar shaped specimens, and the reported Izod impact strength was the mean value of five test specimens.

Tensile testing was carried out on Instron 3345 UTM equipped with a load cell of 50 kN to analyze the mechanical properties (such as tensile modulus, tensile strength and elongation at break) of blends and nanocomposites samples according to ISO 527. The tensile properties of the samples were measured at room temperature using 75 mm x 5 mm x 2 mm dumbbell shaped specimens with a crosshead moving rate of 10 mm.min^{-1} and a span length of 35 mm. The mean value of five test specimens was used to represent the tensile test results.

Differential scanning calorimetry (Discovery series of TA instruments) was used to measure peak melting temperature (T_m), peak crystallization temperature (T_c), melting enthalpy (ΔH_f), crystalliza-tion enthalpy (ΔH_c) and percentage crystallinity (X_c) of PP/CHEE blends and PP/10CHEE/FLG nanocomposites. The analysis was con-ducted in dry nitrogen atmosphere with a flow rate of 50 mL.min^{-1} using about 3-8 mg of the samples weighed in aluminum pans. The samples were subjected to two sets of heating and cooling cycles at a rate of 10 °C.min^{-1}, and the thermograms were generated using the second set of heating and cooling cycles, which included heating from -50 to 200 °C and cooling from 200 to -50 °C.

Thermogravimetric analysis (Discovery series of TA instruments) was used to examine the thermal stability and degradation behavior of the blends and nanocomposites samples. TGA thermograms were recorded by heating about 3-8 mg of the samples under dry nitrogen atmosphere at a heating rate of 10 °C.min^{-1} from 25 to 700 °C. The char yield of the nanocomposite samples was also measured using the thermograms.

Laser flash apparatus Netzsch LFA 447 was used to analyze the thermal conductivity of neat PP, PP/10CHEE blend and graphene nanocomposites as per ASTM E1461. The thermal conductivity of the samples at 25, 50, 75 and 100 °C was measured using disc shaped test specimens (12 mm diameter x 2 mm thickness, with Vespel as a standard) with the aid of Cowan method using the equation:

Thermal conductivity, $k = D\rho_d C_p$

where D is the density, ρ_d is the thermal diffusivity and C_p is the spe-cific heat.

Fourier-transform infrared spectrometer Bruker Tensor II was used to analyze the chemical structure of neat PP, PP/10CHEE blend and FLG nanocomposites. FTIR spectra of the samples were gener-ated using thin sections of the injection molded specimens by collect-ing 32 scans with a resolution of 4 cm^{-1} in the wavenumber range from 4000 to 400 cm^{-1} (transmission mode) with the aid of a diamond attenuated total reflectance (ATR) crystal.

Raman spectrometer Jobin Yvon Horiba LabRAM with a confocal microscope was used to collect the Raman spectra of the samples us-ing a He-Ne laser with an excitation wavelength of 633 nm as the light source. The spectra were generated in a Raman shift range of 500-3500 cm^{-1} using thin injection molded blends and nanocomposites samples at room temperature with the application of a confocal ob-jective at a magnification of 50x.

X'Pert PRO Panalytical powder diffractometer using Cu-Kα radiation (1.5406 Å wavelength) was used to study the wide angle X-ray diffraction behavior of the injection molded blends and nanocomposites samples at room temperature. The X-ray diffractograms of the samples was recorded in the 2θ span of 5-60° with a step rate of 0.017° s^{-1}.

Scanning electron microscope FEI Quanta FEG250 was used to examine fracture micro-morphology of PP/10CHEE blend and its graphene nanocomposites. For this purpose, impact fractured surface of the test specimens was sputter coated with gold.

3.3 Results and Discussion

In this study, PP/CHEE blends and PP/10CHEE/FLG nanocomposites were prepared using direct melt mixing. PP/CHEE blends containing 5, 10, 15 and 20 wt% of copolymer were developed in order to study the influence of copolymer content on the blend performance and to obtain optimal blend composition for the generation of high performance FLG nanocomposites. On the basis of preliminary mechanical characterization, PP/10CHEE blend was used to generate nanocomposites containing 0.5, 1, 3 and 5 wt% of FLG by following the same mixing conditions as used for blends.

Un-notched Izod impact strength of neat PP, PP/CHEE blends and PP/10CHEE/FLG nanocomposites is summarized in Table 3.1. The impact strength of PP was enhanced by ca. 3.5-fold after blending with 5 wt% of CHEE, followed by a decreasing tendency with further increase in CHEE content. These observations indicated the influence of the amount and resultant size of the dispersed CHEE copolymeric phase in the PP/CHEE blends. The enhanced impact strength can be attributed to the crack growth termination mechanism, where CHEE copolymer phase in the blend system serves as an effective stress concentrator and absorbs the impact load, thus, resulting in crazing and shear yielding in the PP matrix, subsequently leading to microvoids. Accordingly, the resultant matrix prevents the growth of cracks by undergoing plastic deformation, thus, exhibiting effective impact energy dissipation [34-36]. Also, the effect of 0.5-5 wt% of FLG on the impact strength of PP/10CHEE blend is summarized in Table 3.1. The impact strength of PP/10CHEE blend increased from 12 to 16 kJ.m^{-2} after the incorporation of 0.5 wt% of FLG, indicating ca. 33% improvement. The PP/10CHEE/FLG nanocomposites containing 1, 3, and 5 wt% FLG exhibited the impact strength values of 12,

Table 3.1 Impact and tensile properties of PP blends and nanocomposites

Sample	Impact strength (KJ/m^2)	Tensile strength (MPa)	Tensile modulus (MPa)	Exten-sion at break (%)
PP	7	39	1011	24
PP/5CHEE	25	40	1222	24
PP/10CHEE	12	40	1348	24
PP/15CHEE	8	39	1337	32
PP/20CHEE	6	39	1312	54
PP/10CHEE/0.5FLG	16	42	1390	13
PP/10CHEE/1FLG	12	44	1437	12
PP/10CHEE/3FLG	9	47	1565	10
PP/10CHEE/5FLG	8	45	1481	9

9, and 8 kJ.m^{-2} respectively, as compared to 7 kJ.m^{-2} for neat PP (Figure 3.1). At low FLG loading, i.e. 0.5 wt%, FLG served as effective crack inhibitor by disturbing the crack propagation path in the PP/10CHEE blend system with convoluted pathways, which correspondingly

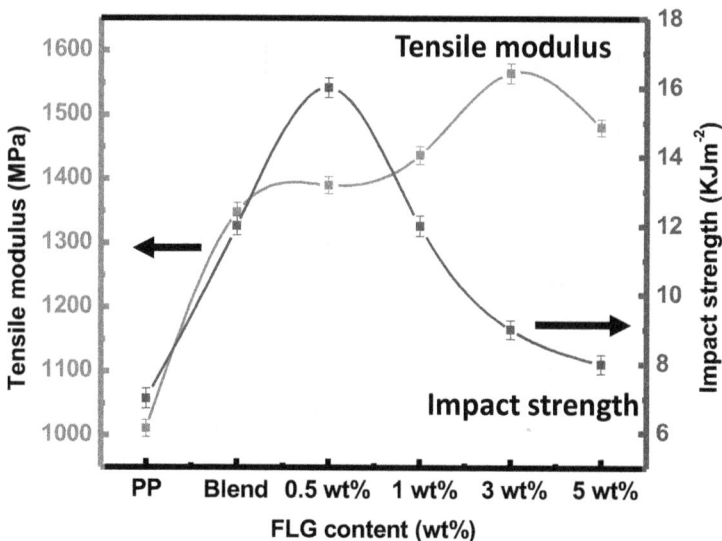

Figure 3.1 Tensile modulus and impact strength of PP, PP/10CHEE blend, and 0.5-5 wt% FLG filled PP/10CHEE/FLG nanocomposites.

enhanced the energy absorption as well as dissipation, thus, leading to enhanced impact performance [28,37,38]. Haghnegahdar *et al.*

[28] also reported an improvement of ca. 23.5% in impact strength for PP/20 wt% EPDM blend nanocomposite with 0.5 wt% FLG. In another study, Parameshwaranpillai *et al.* [31] observed enhanced impact strength values for PP/20 wt% SEBS blend nanocomposites with 0.05-1 wt% graphene nanoplatelets. Table 3.2 provides a comparison of the impact strength observed in this study with various compatibilized PP/graphene nanocomposite systems reported in literature studies.

Table 3.2 Comparison of the impact strength and tensile modulus observed in this study with the compatibilized PP/graphene nanocomposite systems reported in literature

PP nano-composite system	Compatibilizer	inc. in impact str. (%)	opt. filler content (wt%)	inc. in tensile mod. (%)	opt. filler content (wt%)	Ref.
PP/few-layer graphene	Cyclohexylethylene-*co*-ethylene (10 wt%)	33.3	0.5	16.1	3	This work
PP/few-layer graphene	Ethylene propylene diene monomer (20 wt%)	23.5	0.5	17.2	0.5	[28]
PP/multi-layer graphene	Ethylene propylene diene monomer (20 wt%)	7.7	0.5	5.7	0.5	[28]
PP/exfoliated graphene nanoplatelets	Styrene-(ethylene-*co*-butylene)-styrene (20 wt%)	6	1	No effect	0.05-1	[31]
PP/functionalized graphene sheets	Maleic anhydride-*graft*-PP (4.5 wt%)	-	-	27	1.5	[33]

Table 3.1 also lists the tensile properties of neat PP, PP/CHEE blends and PP/CHEE/FLG nanocomposites. Incorporation of 10 wt% CHEE significantly enhanced the tensile modulus, while retaining the tensile strength of PP, followed by a decrease in the properties on further increase in the copolymer content. The observed performance is closely related to the capability of PP and CHEE in the blend system to undergo concurrent deformation during the application of tensile load, which is probably caused by the defect-free blend interface, resulting from the appreciable compatibility and interaction between the blend components [39,40]. Additionally, the elongation at break of neat PP was observed to enhance with CHEE copolymer amount. For instance, a value of 54% was observed for PP blend with 20 wt% CHEE. As expected, the addition of FLG enhanced the tensile modulus and strength of PP in the presence of 10 wt% CHEE. The properties were observed to increase with FLG content up to 3 wt%, followed by a decrease on further increasing the FLG content (Figure 3.1). Specifically, the tensile strength values for 0.5, 1, 3, and 5 wt% FLG containing PP/10CHEE nanocomposites were 42, 44, 47, and 45 MPa respectively in comparison with 39 MPa for neat PP (Figure 3.2). The enhanced tensile modulus and strength of nanocomposites indicated effective FLG dispersion in the PP matrix and strong interfacial interactions [28,41,42]. Table 3.2 also provides a comparison of the

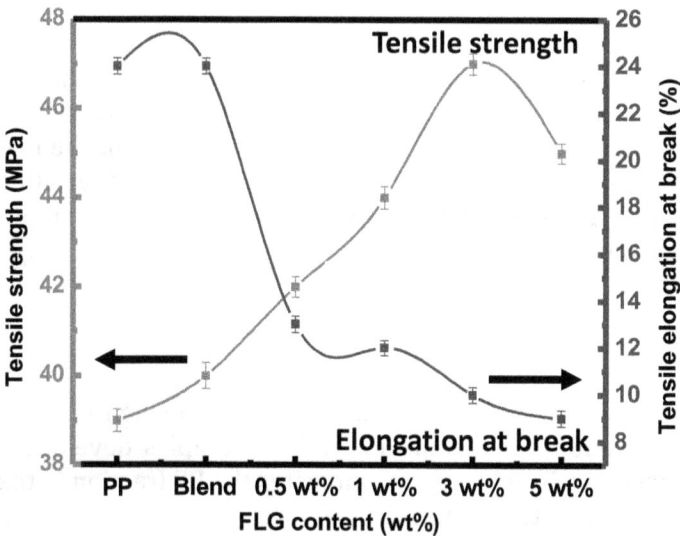

Figure 3.2 Tensile strength and elongation at break of PP, PP/10CHEE blend and 0.5-5 wt% FLG filled PP/10CHEE/FLG nanocomposites.

tensile modulus observed in this study with the compatibilized PP/graphene nanocomposite systems reported in literature. Furthermore, the mechanical performance of PP/10CHEE/FLG nanocomposites was in good agreement with the degree of crystallinity presented in Table 3.3. Ryu *et al.* [13] also reported similar trend for the mechanical performance of the PP/aminated GO nanocomposites.

Table 3.3 Calorimetric properties of neat PP, PP/CHEE blends and PP/10CHEE/FLG nanocomposites

Sample	T_c (°C)	ΔH_c (J/g)	T_m (°C)	ΔH_f (J/g)	X_c (%)	ΔX_c (%)
Neat PP	128	99	168	91	44	-
PP/5CHEE blend	126	96	167	86	44	0
PP/10CHEE blend	125	89	167	83	45	1
PP/15CHEE blend	125	85	167	79	45	1
PP/20CHEE blend	124	80	167	75	45	1
PP/10CHEE/0.5FLG nanocomposite	124	95	168	98	53	9
PP/10CHEE/1FLG nanocomposite	124	93	168	100	54	10
PP/10CHEE/3FLG nanocomposite	124	91	168	99	55	11
PP/10CHEE/5FLG nanocomposite	122	88	169	91	52	8

DSC second heating and cooling thermograms of neat PP, CHEE, PP/CHEE blends and PP/10CHEE/FLG nanocomposites are depicted in Figures 3.3 and 3.4. The calorimetric properties of the materials, such as T_m, T_c, ΔH_f, ΔH_c, X_c and percentage crystallinity variation (ΔX_c), are listed in Table 3.3. The following relationship was employed to calculate X_c of the samples [18,31]:

$$X_c = \frac{\Delta H}{\Delta H_0 w_{pp}} \times 100$$

where ΔH_0 (207.1 J/g [31]) and ΔH represent the melting enthalpy of theoretically fully crystalline PP and PP samples developed in the study respectively, whereas w_{pp} indicates the PP fraction in the blend and nanocomposite samples.

The miscibility of CHEE copolymer with PP was supported by the presence of a single melting transition in the melting thermogram of PP/CHEE blends (Figure 3.3). From Table 3.3, it is clear that the CHEE

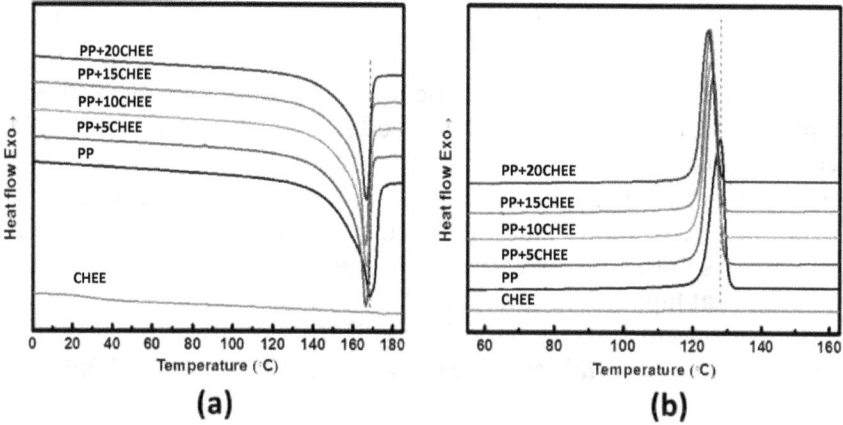

Figure 3.3 DSC (a) melting and (b) crystallization curves of neat PP, CHEE copolymer and 5-20 wt% CHEE containing PP/CHEE blends.

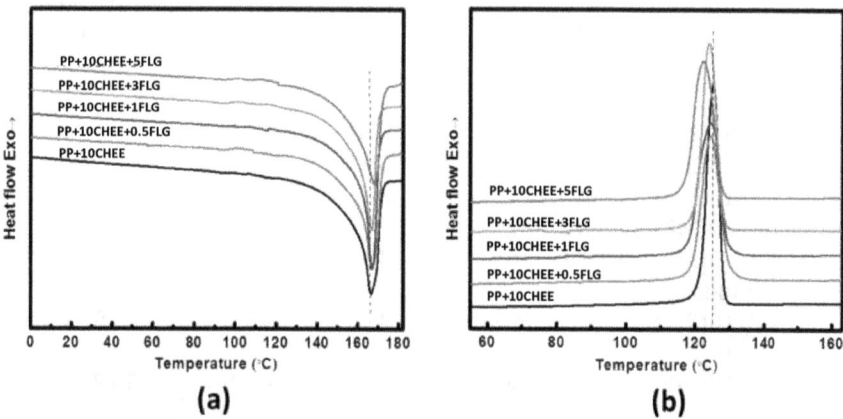

Figure 3.4 DSC (a) melting and (b) crystallization curves of PP/10CHEE blend and 0.5-5 wt% FLG containing PP/10CHEE/FLG nanocomposites.

phase had no significant influence on the T_m of PP (168 °C) [43]. Also, the T_c of the PP/CHEE blends exhibited a decrease of 2-4°C as compared to neat PP (128 °C). Both melting and crystallization enthalpies were observed to decrease with increase in CHEE content in the PP/CHEE blends. The X_c of neat PP and PP phase in PP/CHEE blends was almost consistent, thus, indicating that the presence of CHEE copolymer had no negative effect on the arrangement of PP crystallites [44]. The incorporation of FLG in PP/10CHEE blend had no notable effect on its T_m and T_c. Interestingly, the ΔH_f of PP/10CHEE blend was observed to increase markedly with FLG content till 3 wt%. The ΔH_c values of PP/10CHEE/FLG nanocomposites also displayed almost similar behavior. X_c of PP was observed to strongly enhance with FLG content up to 3 wt%, in the presence of 10 wt% of CHEE copolymer, which indicated the effectiveness of FLG as nucleating agent. However, the decrease in X_c and nucleating activity of FLG for the nanocomposites with 5 wt% FLG resulted due to the propensity of FLG to flocculate at higher fractions [11,32,45-47].

Table 3.4 summarizes the 10% and 50% weight loss temperatures ($T_{10\%}$ and $T_{50\%}$) and char yield at 700 °C for the blend and

Table 3.4 TGA data of neat PP, CHEE, PP/CHEE blends and PP/10CHEE/FLG nanocomposites ($T_{10\%}$: initial degradation temperature and $T_{50\%}$: maximum degradation temperature)

Sample	$T_{10\%}$ (°C)	$T_{50\%}$ (°C)	Char residue at 700 °C (%)
Neat PP	404	447	-
CHEE	403	438	-
PP/5CHEE blend	398	444	-
PP/10CHEE blend	397	443	-
PP/15CHEE blend	382	435	-
PP/20CHEE blend	377	434	-
PP/10CHEE/0.5FLG nanocomposite	423	450	1.03
PP/10CHEE/1FLG nanocomposite	428	455	2.15
PP/10CHEE/3FLG nanocomposite	435	462	5.12
PP/10CHEE/5FLG nanocomposite	440	466	6.58

nanocomposite materials. The TGA thermograms exhibited a single-step degradation behavior for all samples. In nitrogen environment, PP exhibited $T_{10\%}$ and $T_{50\%}$ values of 404 °C and 447°C respectively. The thermal stability of PP was observed to decrease after blending with CHEE copolymer. The incorporation of FLG enhanced the thermal stability of PP/10CHEE blend, as a function of FLG content. For example, the $T_{10\%}$ of PP/10CHEE blend was increased from 397 to 423, 428, 435, and 440 °C on the inclusion of 0.5, 1, 3 and 5 wt% of FLG respectively, owing to the FLG induced increase in the heat transfer efficiency caused by strong PP/10CHEE blend-FLG interactions [11,18,48,49]. The char yield of ca. 1.03, 2.15, 5.12 and 6.58% was observed for the nanocomposites with 0.5, 1, 3 and 5 wt% FLG. Achaby *et al.* [18] also reported single-step degradation behavior for the PP composites with graphene nanosheets. The authors confirmed the capability of graphene nanosheets to appreciably enhance the thermal stability of PP. An increment of ca. 40 °C in $T_{5\%}$ was observed for the composite with 3 wt% graphene nanosheets. In another study, Li *et al.* [48] highlighted the potential of graphene nanoplatelets to enhance the thermal stability of PP. Accordingly, the authors observed ca. 36 °C higher $T_{10\%}$ after the inclusion of 15 wt% graphene nanoplatelets in the PP matrix.

The thermal conductivity of the polymer nanocomposites depends on many factors, such as filler loading and dispersion state, polymer-filler interfacial interactions, etc. [50]. Figure 3.5 illustrates the thermal conduction behavior of neat PP, PP/10CHEE blend and PP/10CHEE/FLG nanocomposites. Neat PP displayed a thermal conductivity of ca. 0.203 W.(m.K)$^{-1}$, with PP/10CHEE blend also exhibiting a similar value. The addition of FLG enhanced the thermal conductivity as a function of FLG content. Table 3.5 also summarizes the thermal conductivity of the materials in the temperature range of 25-100 °C. On addition of 5 wt% FLG, the thermal conductivity of PP/10CHEE blend improved from 0.201 to 0.324 W.(m.K)$^{-1}$, owing to the increased phonon transfer diffusion attributed to the conducting network evolution in the polymer blend matrix [51,52]. These observations are in good agreement with other literature studies on PP/graphene nanocomposites. For example, Song *et al.* [50] reported improved thermal conductivity of ca. 0.396 W.(m.K)$^{-1}$ after the addition of 5 wt% PP grafted graphene in PP. In another study, Kalaitzidou *et al.* [53] observed a thermal conductivity of ca. 0.4 W.(m.K)$^{-1}$ for PP nanocomposite reinforced with 3 vol% of exfoliated graphite platelets.

Structural characterization of FLG, CHEE, PP, PP/10CHEE blend and PP/10CHEE/FLG nanocomposites was carried out using FTIR in transmission mode using thin sections of the injection molded specimens (Figure 3.6). The IR spectrum of FLG demonstrated two absorption bands at 1995 cm^{-1} and 2114 cm^{-1} corresponding to C-H bending of aromatic groups and C≡C stretching of alkyne groups. The FTIR spectra of neat PP displayed absorption peaks corresponding to asymmetric stretching vibration of CH$_3$ groups at 2957 cm^{-1}, symmetric stretching vibration of CH$_2$ groups at 2916 cm^{-1}, symmetric

stretching vibration of CH₃ groups at 2870 cm⁻¹, symmetric stretching
vibration of CH₂ groups at 2837 cm⁻¹, C-H bending vibration at

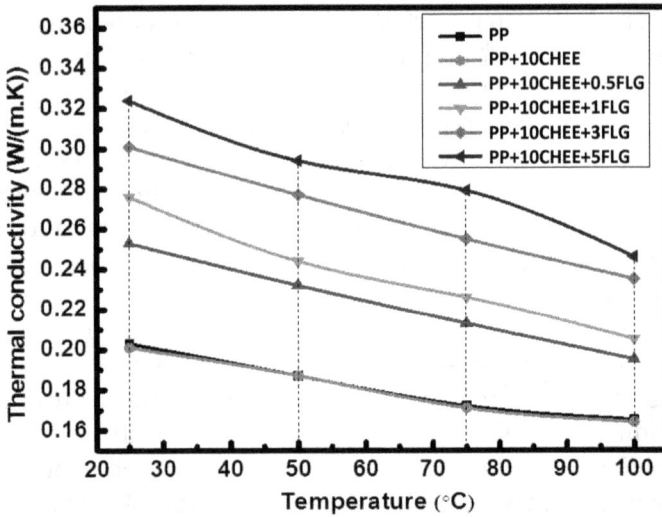

(a)

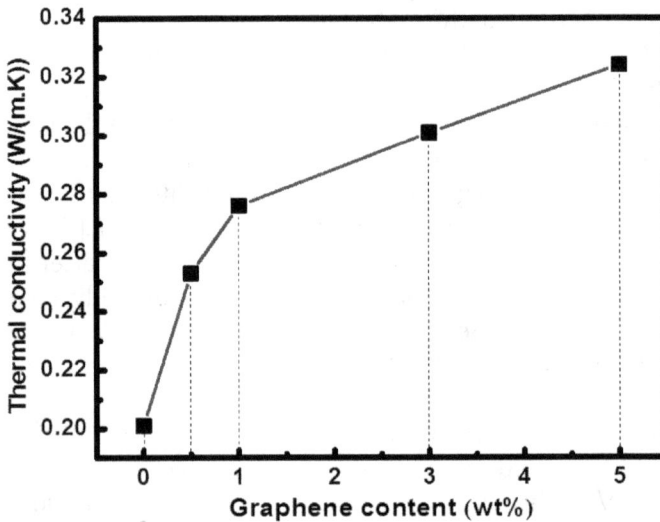

(b)

Figure 3.5 (a) Thermal conductivity variation of PP/10CE70/FLG
nanocomposites with 0.5-5 wt% FLG in temperatures range from 25 to 100
°C and (b) thermal conductivity of PP/10CE70/FLG nanocomposites as a
function of FLG content at 25 °C.

Table 3.5 Thermal conductivity of blends and nanocomposites

Sample	Thermal conductivity (W.(m.K)$^{-1}$)			
	at 25 °C	at 50 °C	at 75 °C	at 100 °C
Neat PP	0.203	0.187	0.172	0.165
PP/10CHEE blend	0.201	0.187	0.171	0.164
PP/10CHEE/0.5FLG	0.253	0.232	0.213	0.195
PP/10CHEE/1FLG	0.276	0.244	0.226	0.205
PP/10CHEE/3FLG	0.301	0.277	0.255	0.235
PP/10CHEE/5FLG	0.324	0.294	0.279	0.246

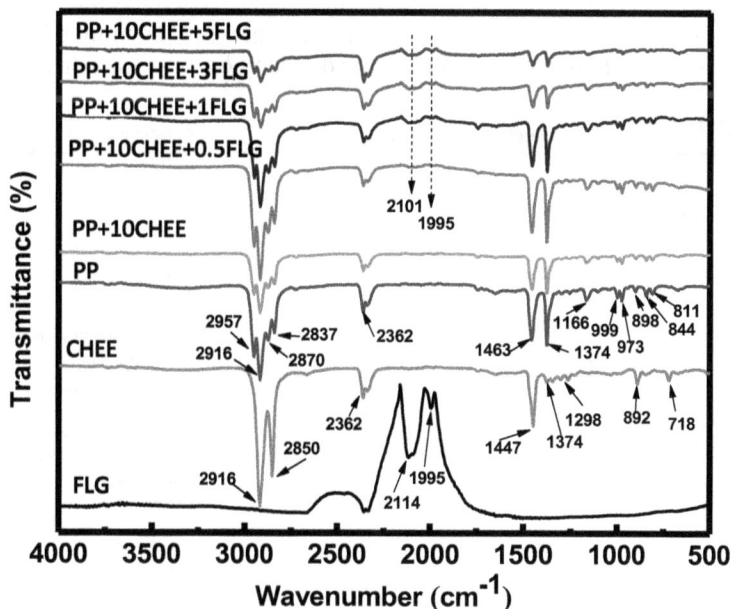

Figure 3.6 FTIR spectra of FLG, CHEE, neat PP, PP/10CHEE blend and PP/10CHEE/FLG nanocomposites.

2362 cm^{-1}, asymmetric bending vibration of CH$_3$ at 1463 cm^{-1}, symmetric bending vibration of CH$_3$ at 1374 cm^{-1}, tertiary methyl skeleton deformation at 999, 973, 844 and 811 cm^{-1} and perpendicular absorption in the stretch direction at 898 cm^{-1} [54]. In the case of CHEE, the characteristic IR peaks were observed for stretching vibration of CH$_2$ at 2916 and 2850 cm^{-1}, bending C-H vibration at 2362 cm^{-1}, bending vibration of CH$_2$ at 1447 cm^{-1}, CH$_2$ wagging at 1374 cm^{-1}, CH$_2$

twisting at 1298 cm^{-1} and CH$_2$ rocking at 892 and 718 cm^{-1} [4,54]. Figure 3.6 revealed the presence of two additional peaks at 2101 and 1995 cm^{-1} in PP/10CHEE/FLG nanocomposites due to FLG. In addition, the peak associated with the C≡C stretching vibration of alkyne at 2114 cm^{-1} on the FLG surface was shifted to 2101 cm^{-1} in the PP/CE-70/FLG nanocomposites, most probably due to its good interaction with the PP/10CHEE blend [55,56]. Zhou *et al.* [56] also reported a shifting of GO characteristic peak at 1703 cm^{-1} to 1690 cm^{-1} in the IR spectra of polypyrrole/GO nanocomposites. Table 3.6 also summarizes the FTIR assignments of FLG, neat PP, CHEE, PP/10CHEE blend and PP/CHEE/FLG nanocomposites.

Table 3.6 Absorption bands of samples in the IR region and their assignments (v: stretching; v_s: symmetric stretching; v_{as}: asymmetric stretching; δ: bending; δ_s: symmetric bending and δ_{as}: asymmetric bending)

Sample	Absorption band (cm^{-1})	Assignment
Neat PP	2957	v_{as} (CH$_3$)
	2916	v_s (CH$_2$)
	2870	v_s (CH$_3$)
	2837	v_s (CH$_2$)
	2362	δ (C-H)
	1463	δ_{as} (CH$_3$)
	1374	δ_s (CH$_3$)
	1166	v_{c-c} of C-CH$_3$
	999	Tertiary methyl skeleton deformation
	973	Tertiary methyl skeleton deformation
	898	Perpendicular absorption in the stretch direction
	844	Tertiary methyl skeleton deformation
	811	Tertiary methyl skeleton deformation
Neat CHEE	2916	v (CH$_2$)
	2850	v (CH$_2$)
	2362	δ (C-H)
	1447	δ (CH$_2$)
	1374	CH$_2$ wagging
	1298	CH$_2$ twisting
	892	CH$_2$ rocking
	718	CH$_2$ rocking
PP/10CHEE blend	All absorption bands of neat PP	
FLG	2114	$v_{C≡C}$ of alkyne group
	1995	δ_{C-H} of aromatic group
PP/10CHEE/FLG nanocomposites	All absorption bands of neat PP, FLG band at 1995, and downshifted FLG band at 2101	

Figure 3.7 illustrates the Raman spectra of FLG, PP/10CHEE blend and PP/10CHEE/FLG nanocomposites. Raman spectrum of FLG

displayed characteristic D, G, 2D and D+G bands at ca. 1348, 1583, 2672 and 2946 cm⁻¹ Raman shift regions respectively. In FLG, D band

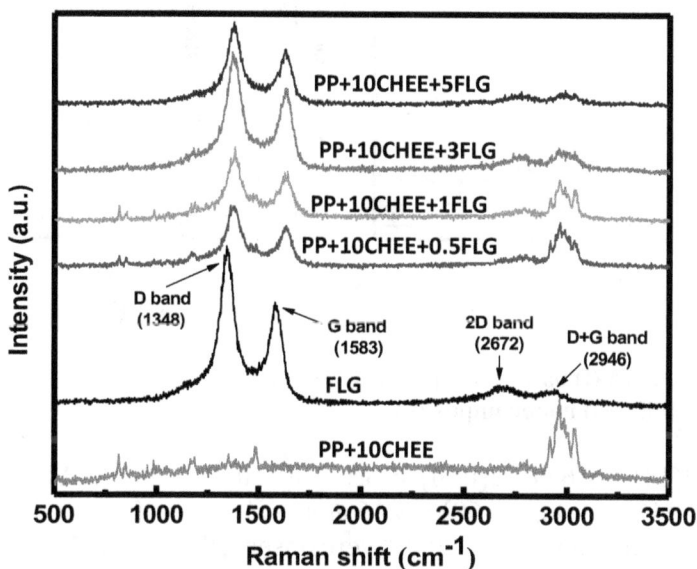

Figure 3.7 Raman spectra of PP/10CHEE blend, FLG and PP/10CHEE/FLG nanocomposites.

associated with the Raman mode inspired C-C ring structural disorders, while G band signified the stretching vibrations of C-C bond. 2D band, also known as second order D band, indicated graphene sheet stacking in the direction of the c-axis [46,57,58]. The Raman spectra of PP/10CHEE/FLG nanocomposites exhibited shifts in the characteristic D, G and 2D bands of FLG as well as merged D+G band of PP/CHEE blend. In addition, the intensity of the characteristic D, G and 2D bands of FLG increased with FLG content up to 3 wt%. However, the intensity of D+G band of PP/10CHEE blend in the 2878-3089 cm⁻¹ Raman shift region was observed to decrease with increasing FLG content.

Wide angle X-ray diffraction patterns of PP/CHEE blends and PP/CHEE/FLG nanocomposites are depicted in Figure 3.8. For neat PP, characteristic diffraction peaks were observed at 14.55°, 17.26°, 19.03°, 21.47°, 22.18°, 25.76° and 29.01° 2θ corresponding to α crystalline planes (110), (040), (130), (111), combinatorial (131) and (041), (160) and (220) respectively [33,46,48,59]. The inter-planar distances of the diffraction peaks associated with the neat polymer

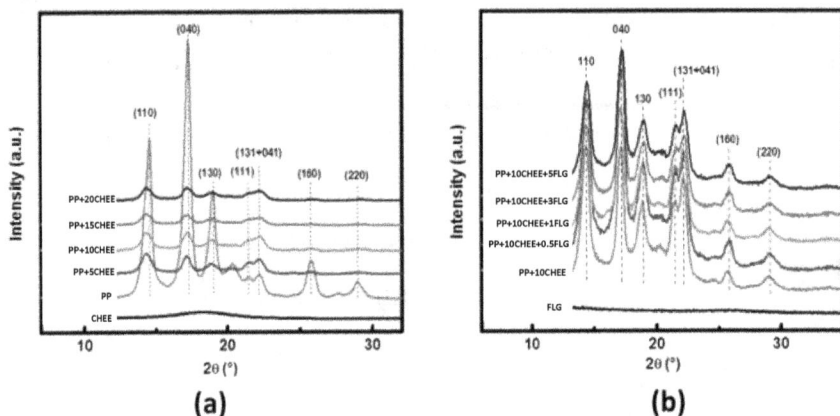

Figure 3.8 WAXD patterns of (a) PP/CHEE blends and (b) PP/10CHEE/FLG nanocomposites.

were d(110) = 6.08, d(040) = 5.13, d(130) = 4.66, d(111) = 4.14, d(131+041) = 4.00, d(160) = 3.46 and d(220) = 3.08 Å. All α crystalline diffraction peaks of neat PP were observed in the WAXD patterns of PP/CHEE blends, however, the peak position of the diffraction peaks corresponding to (110), (040) and (130) planes was shifted to 14.30°, 17.17° and 18.89° 2θ angles, which correspondingly enhanced the inter-planar distances of these peaks to 6.19, 5.16, and 4.69 Å respectively (Table 3.7). The observed changes indicated a

Table 3.7 WAXD data of samples (2θ: diffraction angle and *d*: inter-planar distance)

Sample	Crystalline planes					
	110		040		130	
	2θ (°)	*d* (Å)	2θ (°)	*d* (Å)	2θ (°)	*d* (Å)
Neat PP	14.55	6.08	17.26	5.13	19.03	4.66
PP/CHEE blends	14.30	6.19	17.17	5.16	18.89	4.69
PP/CHEE/FLG nano-composites	14.30	6.19	17.17	5.16	18.89	4.69

good level of compatibility between the blend constituents. The WAXD pattern of FLG did not exhibit any diffraction peak indicating exfoliated morphology and the formation of isolated randomly distributed layered structure [33,60]. In the case of PP/10CHEE/FLG nanocomposites, the X-ray diffraction patterns did not depict any peak associated with FLG, thus, confirming the absence of filler re-stacking during the melt mixing process [28,61]. Also, the diffraction peak positions as well as inter-planar distances in the nanocompo-sites were consistent with PP/10CHEE blend.

SEM images of FLG, as presented in Figure 3.9, confirmed the lay-ered surface morphology. In addition, the micro-morphology of the impact fractured surfaces of neat PP, PP/10CHEE blend and PP/10CHEE/FLG nanocomposites was studied with SEM, as shown in

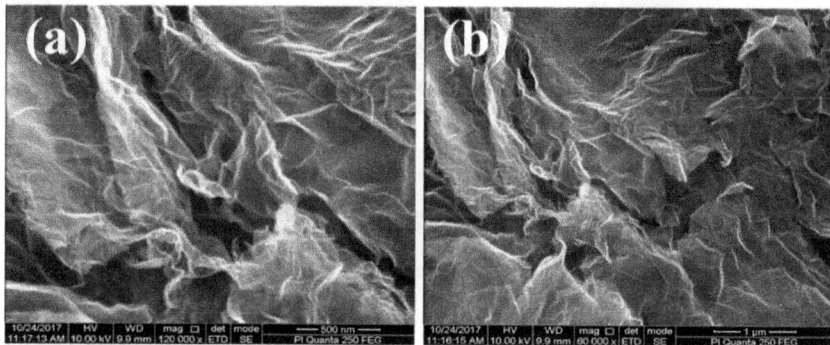

Figure 3.9 SEM images of FLG at (a) high and (b) low magnification.

in Figures 3.10 and 3.11. Higher extent of CHEE dispersed phase pull-outs or microvoids (highlighted circles in SEM images) were ob-served for PP/10CHEE blend (Figure 3.10b) and its nanocomposites with 0.5 and 1 wt% FLG (Figure 3.10c & 3.10d), in comparison with PP/10CHEE/3FLG (Figure 3.10e) and PP/10CHEE/5FLG (Figure 3.10f) nanocomposites. The observed pull-out of the CHEE dispersed phase from the continuous PP matrix indicated debonding or cavita-tion upon application of impact force. According to the crack termi-nation mechanism, the CHEE copolymer phase can function as suc-cessful stress absorber and generates crazing as well as shear yield-ing in the continuous PP matrix, thereby, forming microvoids through the debonding or cavitation at the interface [28]. The morphological observations from the un-notched Izod impact fractured samples were, thus, in good agreement with the observed impact strength be-havior. The fractograph of PP/10CHEE blend in Figure 3.10b revealed effective distribution of CHEE phase in the PP continuous phase, ac-cordingly the impact strength of PP was observed to improve. It could be seen from Figure 3.10c that the incorporation of 0.5 wt% of FLG in PP/10CHEE blend was beneficial to improve the interface between

PP and CHEE copolymer (in terms of void filling), resulting from the effective FLG dispersion. The strengthened interfaces would have

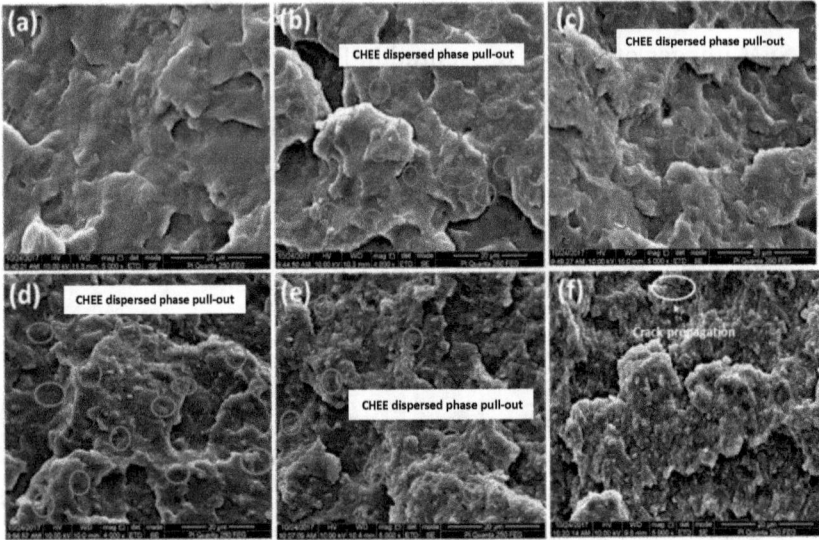

Figure 3.10 SEM micrographs showing the micro-morphology of the impact fractured surfaces of (a) neat PP, (b) PP/10CHEE blend and PP/10CHEE/FLG nanocomposites with FLG content of (c) 0.5, (d) 1, (e) 3 and (f) 5 wt%.

acted as effective crack inhibitors, thus, offering improved impact performance [31,62]. In the case of PP/10CHEE/1FLG nanocomposite (Figure 3.10d), the void filling capability of 1 wt% FLG was not as effective as observed for the 0.5 wt% nanocomposite, which resulted in reduced impact strength. With increase in the FLG content in the PP/10CHEE/FLG nanocomposites, the impact strength was observed to decrease due to the flocculation tendency of FLG [31]. The fractograph of PP/10CHEE/FLG nanocomposite containing 5 wt% FLG also revealed the crack propagation caused by the stress liberation through the FLG flocculates. Additionally, the surface roughness of the PP/10CHEE blend was observed to increase with FLG content due to the flocculation tendency of FLG (Figure 3.11).

3.4 Conclusions

In this study, the successful generation of PP/CHEE blends and PP/10CHEE/graphene nanocomposites using direct melt mixing has

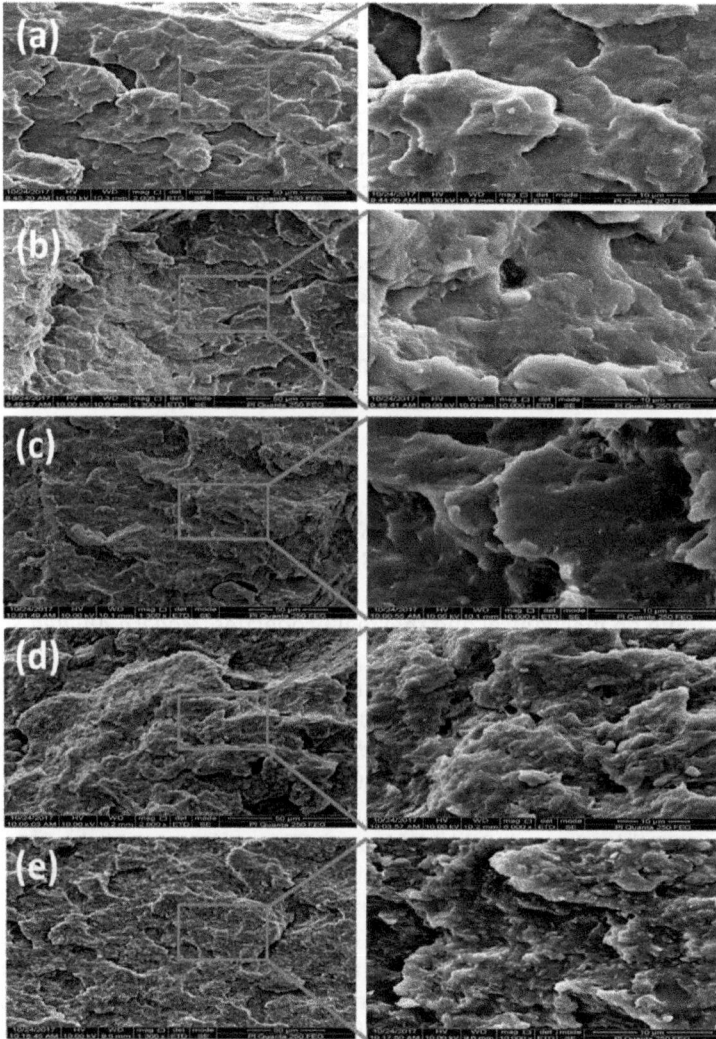

Figure 3.11 SEM images showing the micro-morphology of the impact fractured surfaces of (a) PP/10CHEE blend and PP/10CHEE/FLG nanocomposites with FLG content of (b) 0.5, (c) 1, (d) 3 and (e) 5 wt%.

been demonstrated. Among the various PP/CHEE blend compositions, the blend with 10 wt% CHEE exhibited optimal impact and tensile properties. The incorporation of a low amount of FLG (0.5 wt%) in the nanocomposite formulation was advantageous to enhance the impact performance by ca. 129% due to its capability to act as effective crack inhibitor. Tensile results revealed an optimal enhancement of ca. 55% and 20% in tensile modulus and strength respectively for the nanocomposite with 3 wt% FLG, which was further supported by the observations from DSC. With increasing FLG content in the nano-composites, the heat transfer efficiency was observed to increase, which resulted in enhanced thermal stability. The resultant PP nano-composites also exhibited increasing thermal conductivity with FLG content, and the extent of enhancement was ca. 62% in the case of PP/10CHEE nanocomposite with 5 wt% FLG. The observed findings support the potential use of the developed materials for a wide range of applications in the field of automotive materials, electronic and electrical devices, building and construction materials, among others.

References

1. *Functional Polymer Blends: Synthesis, Properties, and Performance*, Mittal, V. (ed.), CRC Press, USA (2012).
2. Ibrahim, B. A., and Kadum, K. M. (2010) Influence of polymer blend-ing on mechanical and thermal properties. *Modern Applied Science*, **4**, 157-161.
3. Parameswaranpillai, J., Thomas, S., and Grohens, Y. (2014) Polymer blends: state of the art, new challenges, and opportunities. In: *Characterization of Polymer Blends*, Thomas, S., Grohens, Y., and Jyotish-kumar, P. (eds.), Wiley-VCH, Germany, pp. 1-6.
4. Mittal, V., Luckachan, G. E., and Matsko, N. B. (2014) PE/chlorinated-PE blends and PE/chlorinated-PE/graphene oxide nanocompo-sites: Morphology, phase miscibility, and interfacial interactions. *Macromolecular Chemistry and Physics*, **215**, 255-268.
5. Hong, C. H., Lee, Y. B., Bae, J. W., Jho, J. Y., Nam, B. U., and Hwang, T. W. (2005) Preparation and mechanical properties of polypropyl-ene/clay nanocomposites for automotive parts application. *Journal of Applied Polymer Science*, **98**, 427-433.
6. Ramos Filho, F. G., Mélo, T. J. A., Rabello, M. S., and Silva, S. M. (2005) Thermal stability of nanocomposites based on polypropylene and bentonite. *Polymer Degradation and Stability*, **89**, 383-392.
7. Fu, X., Yao, C., and Yang, G. (2015) Recent advances in graphene/pol-yamide 6 composites: a review. *RSC Advances*, **5**, 61688-61702.
8. Müller, K., Bugnicourt, E., Latorre, M., Jorda, M., Echegoyen Sanz, Y., Lagaron, J. M., Miesbauer, O., Bianchin, A., Hankin, S., and Bölz, U. (2017) Review on the processing and properties of polymer nano-composites and nanocoatings and their applications in the packag-ing, automotive and solar energy fields. *Nanomaterials*, **7**, 74.
9. Varghese, A. M., Rangaraj, V. M., Mun, S. K., Macosko, C. W., and Mit-tal, V. (2018) Effect of graphene on polypropylene/maleic

anhydride-graft-ethylene-vinyl acetate (PP/EVA-g-MA) blend: Mechanical, thermal, morphological, and rheological properties. *Industrial & Engineering Chemistry Research*, **57**, 7834-7845.

10. Alexandre, M., and Dubois, P. (2000) Polymer-layered silicate nanocomposites: preparation, properties and uses of a new class of materials. *Materials Science and Engineering, R: Reports*, **28**, 1-63.

11. Mittal, V., and Chaudhry, A. U. (2016) Polyethylene-thermally reduced graphene nanocomposites: comparison of masterbatch and direct melt mixing approaches on mechanical, thermal, rheological, and morphological properties. *Colloid and Polymer Science*, **294**, 1659-1670.

12. Vallés, C., Abdelkader, A. M., Young, R. J., and Kinloch, I. A. (2014) Few layer graphene-polypropylene nanocomposites: the role of flake diameter. *Faraday Discussions*, **173**, 379-390.

13. Ryu, S. H., and Shanmugharaj, A. (2014) Influence of hexamethylene diamine functionalized graphene oxide on the melt crystallization and properties of polypropylene nanocomposites. *Materials Chemistry and Physics*, **146**, 478-486.

14. Mukhopadhyay, P., and Gupta, R. K. (2011) Trends and frontiers in graphene-based polymer nanocomposites. *Plastics Engineering*, **67**, 32-42.

15. Mittal, V., and Chaudhry, A. U. (2015) Polymer–graphene nanocomposites: effect of polymer matrix and filler amount on properties. *Macromolecular Materials and Engineering*, **300**, 510-521.

16. Park, S., and Ruoff, R. S. (2015) Synthesis and characterization of chemically modified graphenes. *Current Opinion in Colloid & Interface Science*, **20**, 322-328.

17. Kuilla, T., Bhadra, S., Yao, D., Kim, N. H., Bose, S., and Lee, J. H. (2010) Recent advances in graphene based polymer composites. *Progress in Polymer Science*, **35**, 1350-1375.

18. El Achaby, M., Arrakhiz, F. E., Vaudreuil, S., el Kacem Qaiss, A., Bousmina, M., and Fassi-Fehri, O. (2012) Mechanical, thermal, and rheological properties of graphene-based polypropylene nanocomposites prepared by melt mixing. *Polymer Composites*, **33**, 733-744.

19. Mittal, V., Chaudhry, A. U., and Luckachan, G. E. (2014) Biopolymer-thermally reduced graphene nanocomposites: Structural characterization and properties. *Materials Chemistry and Physics*, **147**, 319-332.

20. *Graphite, Graphene, and their Polymer Nanocomposites*, Mukhopadhyay, P., and Gupta, R. K. (eds.), CRC Press, USA (2012).

21. Mittal, V., and Chaudhry, A. U. (2015) Effect of amphiphilic compatibilizers on the filler dispersion and properties of polyethylene - thermally reduced graphene nanocomposites. *Journal of Applied Polymer Science*, **132**, doi: 10.1002/app.42484.

22. Chaudhry, A. U., and Mittal, V. (2015) Masterbatch approach to gen-

erate HDPE/CPE/graphene nanocomposites. In: *Synthesis Techniques for Polymer Nanocomposites*, Mittal, V. (ed.), Wiley-VCH, Germany, pp. 31-50.

23. Wang, Y., Wang, J., and Chen, S. (2014) Role of surfactant molecular weight on morphology and properties of functionalized graphite oxide filled polypropylene nanocomposites. *Express Polymer Letters*, **8**, 881-894.

24. Bikiaris, D., Vassiliou, A., Chrissafis, K., Paraskevopoulos, K., Jannakoudakis, A., and Docoslis, A. (2008) Effect of acid treated multi-walled carbon nanotubes on the mechanical, permeability, thermal properties and thermo-oxidative stability of isotactic polypropylene. *Polymer Degradation and Stability*, **93**, 952-967.

25. Dai Lam, T., Hoang, T. V., Quang, D. T., and Kim, J. S. (2009) Effect of nanosized and surface-modified precipitated calcium carbonate on properties of CaCO 3/polypropylene nanocomposites. *Materials Science and Engineering A*, **501**, 87-93.

26. Lonkar, S. P., Morlat-Therias, S., Caperaa, N., Leroux, F., Gardette, J.-L., and Singh, R. (2009) Preparation and nonisothermal crystallization behavior of polypropylene/layered double hydroxide nanocomposites. *Polymer*, **50**, 1505-1515.

27. Wang, Y., and Tsai, H. B. (2012) Thermal, dynamic-mechanical, and dielectric properties of surfactant intercalated graphite oxide filled maleated polypropylene nanocomposites. *Journal of Applied Polymer Science*, **123**, 3154-3163.

28. Haghnegahdar, M., Naderi, G., and Ghoreishy, M. (2017) Fracture toughness and deformation mechanism of un-vulcanized and dynamically vulcanized polypropylene/ethylene propylene diene monomer/graphene nanocomposites. *Composites Science And Technology*, **141**, 83-98.

29. Haghnegahdar, M., Naderi, G., and Ghoreishy, M. (2017) Electrical and thermal properties of a thermoplastic elastomer nanocomposite based on polypropylene/ethylene propylene diene monomer/graphene. *Soft Materials*, **15**, 82-94.

30. Haghnegahdar, M., Naderi, G., and Ghoreishy, M. (2017) Microstructure and mechanical properties of nanocomposite based on polypropylene/ethylene propylene diene monomer/graphene. *International Polymer Processing*, **32**, 72-83.

31. Parameswaranpillai, J., Joseph, G., Shinu, K., Sreejesh, P., Jose, S., Salim, N. V., and Hameed, N. (2015) The role of SEBS in tailoring the interface between the polymer matrix and exfoliated graphene nanoplatelets in hybrid composites. *Materials Chemistry and Physics*, **163**, 182-189.

32. Parameswaranpillai, J., Joseph, G., Shinu, K., Jose, S., Salim, N. V., and Hameed, N. (2015) Development of hybrid composites for automotive applications: effect of addition of SEBS on the morphology,

mechanical, viscoelastic, crystallization and thermal degradation properties of PP/PS–x GnP composites. *RSC Advances*, **5**, 25634-25641.

33. Li, C.-Q., Zha, J.-W., Long, H.-Q., Wang, S.-J., Zhang, D.-L., and Dang, Z.-M. (2017) Mechanical and dielectric properties of graphene incorporated polypropylene nanocomposites using polypropylene-graft-maleic anhydride as a compatibilizer. *Composites Science and Technology*, **153**, 111-118.

34. Liu, G., and Qiu, G. (2013) Study on the mechanical and morphological properties of toughened polypropylene blends for automobile bumpers. *Polymer Bulletin*, **70**, 849-857.

35. Panda, B. P., Mohanty, S., and Nayak, S. K. (2015) Mechanism of toughening in rubber toughened polyolefin - A review. *Polymer-Plastics Technology and Engineering*, **54**, 462-473.

36. Bu, H., Qiu, W., Tan, Z., Li, Q., and Rong, J. (2015) Study on toughening of poly (4-methyl-1-pentene) with various thermoplastic elastomers. *Journal of Thermoplastic Composite Materials*, **28**, 1334-1342.

37. Fu, S., Li, N., Wang, K., Zhang, Q., and Fu, Q. (2015) Reduction of graphene oxide with the presence of polypropylene micro-latex for facile preparation of polypropylene/graphene nanosheet composites. *Colloid and Polymer Science*, **293**, 1495-1503.

38. Shokrieh, M., and Joneidi, V. (2015) Characterization and simulation of impact behavior of graphene/polypropylene nanocomposites using a novel strain rate–dependent micromechanics model. *Journal of Composite Materials*, **49**, 2317-2328.

39. Zhu, L., Fan, H.-N., Yang, Z.-Q., and Xu, X.-H. (2010) Evaluation of phase morphology, rheological, and mechanical properties based on polypropylene toughened with poly (ethylene-co-octene) *Polymer-Plastics Technology and Engineering*, **49**, 208-217.

40. Contreras, V., Cafiero, M., Da Silva, S., Rosales, C., Perera, R., and Matos, M. (2006) Characterization and tensile properties of ternary blends with PA-6 nanocomposites. *Polymer Engineering & Science*, **46**, 1111-1120.

41. Chammingkwan, P., Matsushita, K., Taniike, T., and Terano, M. (2016) Enhancement in mechanical and electrical properties of polypropylene using graphene oxide grafted with end-functionalized polypropylene. *Materials*, **9**, 240.

42. Yuan, B., Bao, C., Song, L., Hong, N., Liew, K. M., and Hu, Y. (2014) Preparation of functionalized graphene oxide/polypropylene nanocomposite with significantly improved thermal stability and studies on the crystallization behavior and mechanical properties. *Chemical Engineering Journal*, **237**, 411-420.

43. Ferrer, G. G., Sánchez, M. S., Sánchez, E. V., Colomer, F. R., and Ribelles, J. L. G. (2000) Blends of styrene–butadiene–styrene triblock

copolymer and isotactic polypropylene: morphology and thermo-mechanical properties. *Polymer International*, **49**, 853-859.

44. Liao, C. Z., and Tjong, S. C. (2011) Effects of carbon nanofibers on the fracture, mechanical, and thermal properties of PP/SEBS-g-MA blends. *Polymer Engineering & Science*, **51**, 948-958.

45. Milani, M. A., González, D., Quijada, R., Basso, N. R., Cerrada, M. L., Azambuja, D.S., and Galland, G.B. (2013) Polypropylene/graphene nanosheet nanocomposites by in situ polymerization: synthesis, characterization and fundamental properties. *Composites Science and Technology*, **84**, 1-7.

46. Zhao, S., Chen, F., Zhao, C., Huang, Y., Dong, J.-Y., and Han, C. C. (2013) Interpenetrating network formation in isotactic polypropyl-ene/graphene composites. *Polymer*, **54**, 3680-3690.

47. Triantou, M. I., Stathi, K. I., and Tarantili, P. A. (2014) Rheological and thermomechanical properties of graphene/ABS/PP nanocom-posites. World Academy of Science, Engineering and Technology, *International Journal of Chemical, Molecular, Nuclear, Materials and Metallurgical Engineering*, **8**, 967-972.

48. Li, Y., Zhu, J., Wei, S., Ryu, J., Sun, L., and Guo, Z. (2011) Poly (propyl-ene)/graphene nanoplatelet nanocomposites: melt rheological be-havior and thermal, electrical, and electronic properties. *Macromo-lecular Chemistry and Physics*, **212**, 1951-1959.

49. Chatterjee, A., and Deopura, B. (2006) Thermal stability of polypro-pylene/carbon nanofiber composite. *Journal of Applied Polymer Sci-ence*, **100**, 3574-3578.

50. Song, P., Cao, Z., Cai, Y., Zhao, L., Fang, Z., and Fu, S. (2011) Fabrica-tion of exfoliated graphene-based polypropylene nanocomposites with enhanced mechanical and thermal properties. *Polymer*, **52**, 4001-4010.

51. Inuwa, I., Hassan, A., and Shamsudin, S. (2014) Thermal properties, structure and morphology of graphene reinforced polyethylene ter-ephthalate/polypropylene nano composites. *The Malaysian Journal of Analytical Sciences*, **18**, 466-477.

52. Huang, X., Zhi, C., and Jiang, P. (2012) Toward effective synergetic effects from graphene nanoplatelets and carbon nanotubes on ther-mal conductivity of ultrahigh volume fraction nanocarbon epoxy composites. *The Journal of Physical Chemistry C*, **116**, 23812-23820.

53. Kalaitzidou, K., Fukushima, H., and Drzal, L. T. (2007) Multifunc-tional polypropylene composites produced by incorporation of ex-foliated graphite nanoplatelets. *Carbon*, **45**, 1446-1452.

54. Deshmane, C., Yuan, Q., Perkins, R., and Misra, R. (2007) On striking variation in impact toughness of polyethylene-clay and polypropyl-ene–clay nanocomposite systems: the effect of clay-polymer inter-action. *Materials Science and Engineering: A*, **458**, 150-157.

55. Wen, X., Wang, Y., Gong, J., Liu, J., Tian, N., Wang, Y., Jiang, Z., Qiu, J.,

and Tang, T. (2012) Thermal and flammability properties of polypropylene/carbon black nanocomposites. *Polymer Degradation and Stability*, **97**, 793-801.

56. Zhou, H., Han, G., Xiao, Y., Chang, Y., and Zhai, H.-J. (2014) Facile preparation of polypyrrole/graphene oxide nanocomposites with large areal capacitance using electrochemical codeposition for supercapacitors. *Journal of Power Sources*, **263**, 259-267.

57. Ryu, S. H., and Shanmugharaj, A. (2014) Influence of long-chain alkylamine-modified graphene oxide on the crystallization, mechanical and electrical properties of isotactic polypropylene nanocomposites. *Chemical Engineering Journal*, **244**, 552-560.

58. Pimenta, M., Dresselhaus, G., Dresselhaus, M. S., Cancado, L., Jorio, A., and Saito, R. (2007) Studying disorder in graphite-based systems by Raman spectroscopy. *Physical Chemistry Chemical Physics*, **9**, 1276-1290.

59. Gopakumar, T., and Page, D. (2004) Polypropylene/graphite nanocomposites by thermo-kinetic mixing. *Polymer Engineering & Science*, **44**, 1162-1169.

60. Kim, H., and Macosko, C. W. (2008) Morphology and properties of polyester/exfoliated graphite nanocomposites. *Macromolecules*, **41**, 3317-3327.

61. Liang, J., Huang, Y., Zhang, L., Wang, Y., Ma, Y., Guo, T., and Chen, Y. (2009) Molecular-level dispersion of graphene into poly (vinyl alcohol) and effective reinforcement of their nanocomposites. *Advanced Functional Materials*, **19**, 2297-2302.

62. He, S., Petkovich, N. D., Liu, K., Qian, Y., Macosko, C. W., and Stein, A. (2017) Unsaturated polyester resin toughening with very low loadings of GO derivatives. *Polymer*, **110**, 149-157.

4

Polymers and Composites for Optical Applications

4.1 Introduction

The optical properties of a polymer are largely dependent on the crystallinity of the material, type of polymer or copolymer, additives and fillers used for the enhancement of certain properties, etc. [1]. The optical properties of polymers are further affected by the mode of synthesis and synthesis conditions, external environment, temperature and pressure of the surroundings, etc. The optical properties of the polymers are also affected by aging even if the other physical and mechanical properties are intact [2]. There are various attributes of the polymers' optical properties such as refractive index, birefringence, Abbe number, polarization, transparency, reflection, etc. Overall, the optical properties of polymers are one of the main factors driving their application in various devices like lasers, wave guides, wave grating, lenses, etc. [3]. The interaction of polymer matrices with light, leading to the change in electronic band structure, is responsible for the most of the optical phenomena. In addition, other interactions are also observed to impart optical characteristics to the polymers, such as the interaction between an external magnetic field and the electronic structure giving rise to magneto-optic properties [4,5]. This chapter presents a comprehensive review of the various optical properties of the polymeric materials as well as different polymers applied in various optical applications.

4.2 Terms Related to the Optical Properties of Polymers

Refractive index can be defined as the ratio of the velocity of light in a particular medium compared to the velocity in vacuum. The Lorenz-Lorentz equation provides the relationship between the various parameters of a polymer, including the refractive index, expressed as follows:

Liyamol Jacob and Vikas Mittal*, The Petroleum Institute (part of Khalifa University of Science and Technology), Abu Dhabi, UAE
*Current address: Bletchington, Wellington County, Australia
© 2019 Central West Publishing, Australia

$$\frac{n^2-1}{n^2+2} = 4\frac{4\pi\rho Na}{3\,Mw}\alpha = \frac{[R]}{Vo}$$

where, n is the refractive index, ρ is the density, N_A is the Avogadro number, M_w represents the molecular weight, α is the linear molecular polarizability, R is the molar refraction and Vo represents the molecular volume of the polymer repeating unit. These factors can be adjusted to tune the refractive index of the polymer matrix [6].

Abbe number indicates the dispersion of a material, and it can be expressed by the following equation:

$$V = \frac{N_{D-1}}{N_{F-N_C}}$$

where, V is the Abbe number and N_D, N_F and N_C are the refractive indices of the material at the wavelengths of the Fraunhofer D-, F- and C- spectral lines [7].

Birefringence is the optical property where the refraction of a material is dependent on the direction of propagation and polarization of light. Under mechanical stress, most of the polymeric materials are birefringent [8,9].

Transparency indicates the ability of a material, to allow the light to pass through without scattering. In polymers, it is dependent on the fillers and their dispersion state [10].

4.3 High Refractive Index Polymers (HRIPs)

Generation of highly functional materials with attractive optical properties has been a focus of research in recent years [11]. Compared to the inorganic materials used for various optical applications like wave guides, prisms, lenses, etc., polymers with high Abbe number as well as high refractive index have received more attention because of their proccessability, impact resistance, lightweight, ease of dyeing, etc. [12,13]. High refractive index polymers are the polymers which have a refractive index value >1.5. For instance, cast molded thermoset polymers derived from polyisocyanates, polythiols and episulfides have been widely used as eye glasses commercially [14-16]. Projector, camera and pick up lenses made from thermoplastic cyclo-olefin polycarbonate and poly(methyl methacrylate) (PMMA) polymers have also found immense commercial uses [17]. Many advanced optical and optoelectronic devices like complementary metal oxide semiconductor (CMOS) image sensors, micro-lens components for charge coupled devices (CCD), optical adhesives or encapsulants for antireflective coatings, high performance substrates for display devices, etc., require polymers with high refractive index as well as high transparency [18-27].

Intrinsic HRIPs can be defined as polymers which contain a functional group or an atom with high refractive index [28]. C≡C, C, C=C, O (in OH), O (in C=O), phenyl, O (in ether), naphthyl, Cl, S (S–H), Br, S (S–S), I and PH_3 are the most commonly used atoms and functional groups in this respect. Moreover, metallic and π-conjugated systems are also employed for enhancing the refractive index of a given polymer. Intrinsic HRIPs can be synthesized using a number of methods like radical polymerization, polycondensation, Michael polyaddition, step growth polymerization, etc. [29].

4.3.1 Sulfur Containing HRIPs

According to the Lorenz-Lorentz equation, the refractive index of any given sulfur containing polymer is affected by a number of factors like the molecular volume, aromatic content, sulfur content in the repeating unit, etc. [30]. Various recent research studies have reported the development of sulfur containing HRIPs. Polycondensation reaction of two alicyclic dianhydrides 1,2,4,5-cyclohexanetetracarboxylic dianhydride (CHDA) and 1,2,3,4-cyclobutanetetracarboxylic dianhydride (CBDA) with sulphur containing aromatic diamines 2,7-bis(4-aminophenylenesulfanyl)thianthrene (APTT) and 4,4'-thiobis[(p-phenylenesulfanyl)aniline] (3SDA) respectively resulted in a series of semi-alicyclic polyimides (PIs) [31]. The polymerization reaction was performed in two steps using poly(amic acid) (PAA) precursors. The resulting polymers were observed to be flexible as well as tough. In air and nitrogen atmospheres, the polymers showed significant thermal oxidative stability and minimal weight loss till 400 °C. The glass transition temperature range of the developed PIs was 236.5 to 274.1 °C. At 632.8 nm, the PI films exhibited refractive indices in the range of 1.6799-1.7130. The films showed high optical transparency, with a cut-off wavelength less than 350 nm. The polymers were also modified with TiO_2 nanoparticles. At 632.8 nm, the TiO_2 incorporated polymer film demonstrated a higher refractive index of 1.8100. Figure 4.1 shows the schematic of the composite preparation.

Inverse vulcanization of sulfur has also been used to synthesize polymers with high refractive index between 1.75 and 1.86 [32]. The

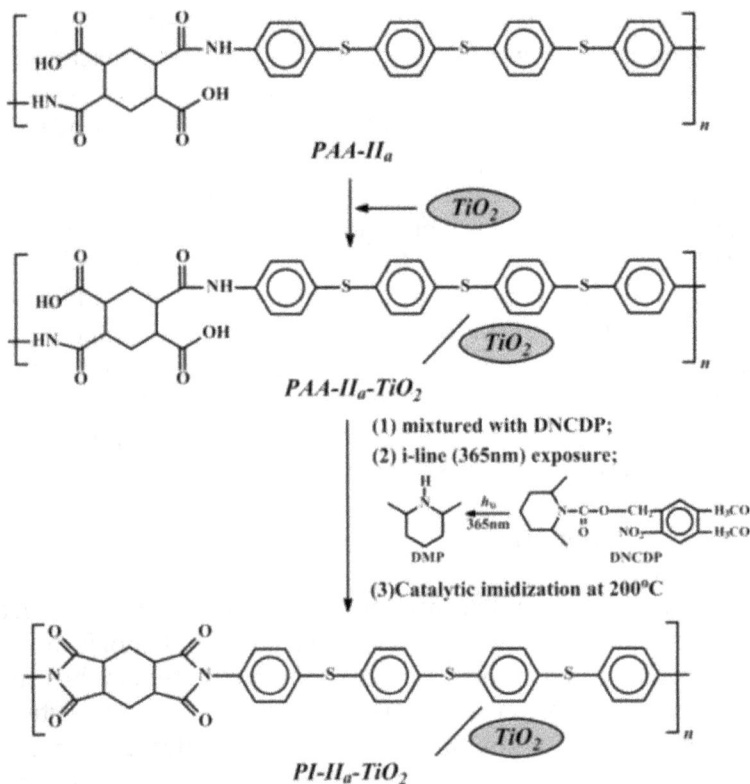

Figure 4.1 Schematic of the composite preparation. Reproduced from Reference 31 with permission from American Chemical Society.

developed materials were suitable to generate lenses for high quality imaging in the near and mid-infra-red regions. In another study, radical polymerization of sulfur containing aliphatic methacrylates or alicyclic methacrylates was reported [33]. For application in optical lenses, high Abbe number as well as high refractive index polymers were also synthesized from methacrylate, norbornene and cyclic dithiocarbonate units [34]. The polymers exhibited stability at high temperatures in air and nitrogen atmospheres. The glass transition behavior was observed between 44-109 °C, whereas the Abbe number was noted between 40.5 and 44.5. The refractive indices were observed in the range 1.592-1.640 at 589 nm. At 400-800 nm, polymers synthesized using spin casting method also exhibited high transparency. In another study, high refractive aromatic polyimides (PIs) containing thianthrene-2,7-disulfanyl moiety were prepared [35]. Two step condensation reaction was employed to synthesize the PI polymer, using aromatic dianhydrides and aromatic diamine 2,7-bis (4-aminophenylenesulfanyl) thianthrene (APTT) as precursors. The films were amorphous, transparent, tough and flexible. The glass transition temperature of the developed polymers was >200 °C. In nitrogen, the polymers demonstrated only 10% weight loss at >500 °C, indicating high thermal stability. At 632.8 nm, the observed refractive indices were >1.73.

4.3.2 Phosphorus Containing HRIPs

The electronic structure of phosphorus gives it high polarizability comparable to nitrogen. The greater polarizability further contributes to a higher refractive index. The small energy band gap in phosphorus makes it a suitable material for HRIPs. It also has the benefit of high transmission ability in visible light region [30]. Significant research focus has been devoted for the development of phosphorus containing optical polymers. Most of the reported studies have also demonstrated the additional benefit of flame retardancy in such polymers. Fujita *et al.* [36] synthesized composites of PMMA with TiO_2 nanoparticles (NPs) modified with oleyl phosphate for various optical applications. The surface modification of TiO_2 nanoparticles using oleyl phosphate was observed to form strong covalent Ti-O-P bonds. The median size of nanoparticles was 16.2 nm, which also helped to prevent Rayleigh scattering. Even with 20% TiO_2 content, the films exhibited excelled transparency in visible light medium. The film with 20% TiO_2 content showed the highest refractive index of 1.86. The refractive index was also observed to be a function of TiO_2 content in the composites.

In another study, polycondensation reaction of diols and phosphonic dichlorides was used to synthesize polyphosphonates with aromatic backbones [37]. The method helped in improving the refractive index of the developed polymers. The termination of the polymer with different alcohols resulted in a tunable glass transition temperatures ranging from 41 to 214 °C. The polymers demonstrated excellent thermal stability up to 450 °C. The highest refractive index was observed to be 1.66. The high Abbe number (>22) confirmed the usefulness of the developed polymers for various optical applications. In another recent study, an epoxy resin were synthesized using a novel curing agent containing phosphorus and sulfur, tris(2-mercaptoethyl) phosphate (TMEP), for LED packaging applications [38]. The resin had good flame retardant properties, with high refractive index of 1.593. The precursors used for the synthesis of the curing agent were sodium hydrosulfide and tris(2-chloroethyl) phosphate (TCEP).

Takahashi *et al.* [39] reported the preparation of OP_TiO$_2$/cyclo-olefin polymer (COP) hybrid films using oleyl-phosphate-modified TiO$_2$ nanoparticles. The surface modification was achieved using phase transfer of TiO$_2$ nanoparticles dispersed in an aqueous solution to the toluene phase containing oleyl phosphate. At 633 nm, the films with 19.1 vol% TiO$_2$ loading exhibited high transmittance of 99.8%. The material with 19.1 vol% TiO$_2$ loading also demonstrated the highest refractive index of 1.83 among all composites. Figure 4.2 shows the variation of refractive index of the polymer with varying TiO$_2$ content.

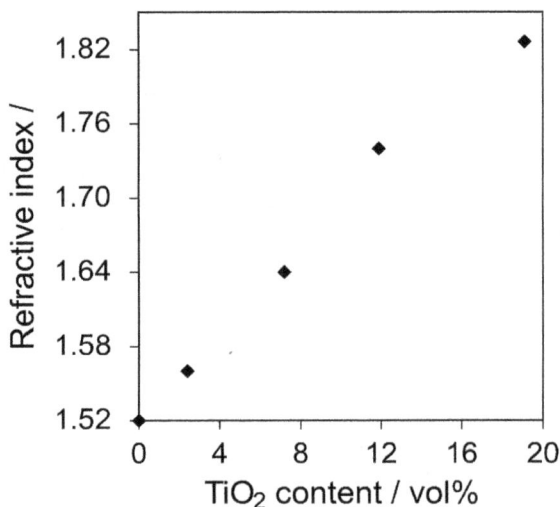

Figure 4.2 Variation in the refractive index of hybrid films as a function of TiO$_2$ content. Reproduced from Reference 39 with permission from American Chemical Society.

Tan *et al.* [40] used thiol-ene photopolymerization to synthesize crosslinked polymer materials containing phosphorus. For this purpose, multifunctional and eco-friendly raw material tris(allyloxymethyl)phosphine oxide (TAOPO) was used as a monomer. Under UV radiation, the reaction of different polythiols with TAOPO resulted in crosslinked poly(phosphine oxide) networks. Phosphorus-carbon bond in TAOPO provided the crosslinked polymers long-term hydrolysis resistance. The polymer exhibited low dielectric loss as well as high dielectric constant, gel content, Abbe number and refractive index, along with an excellent transparency. The cured poly(phosphine oxide) also demonstrated high thermal stability as confirmed by dynamic mechanical analysis (DMA) and thermogravimetric analysis (TGA).

4.3.3 Halogen Containing HRIPs

Halogens are capable of increasing the refractive index of polymers, except fluorine. Fluorine is electronegative and non-polarizable, which reduces the refractive index of the polymeric materials. In other words, the refractive index is directly related to the polarizability of the halogen, i.e. I>Br>Cl. Thus, by tuning the halogen content, it is possible to tune the optical properties of polymers for specific applications. Linkers used during the synthesis of halogen containing polymers also affect the properties of the resultant materials. For example, the glass transition temperature, melting temperature and refractive index reduce with the use of long chain linkers [30].

In a recent study, iodine doped PMMA hybrids were studied for their optical properties [41]. The composites had high refractive index, which was also related to the annealing temperature. At low iodine concentration, the refractive indices were observed to vary non-linearly with extent of doping. In another study, 1-haloalkyne was used to synthesize conjugated polydiynes using an organic reaction route in the presence of potassium iodide [42]. The method was particularly helpful as it involved transition metal-free polymerization. The method also provided benefits of enhanced material performance, decreased metal residues and low synthesis cost. The materials were synthesized at 120 °C under nitrogen, by the polymerization of 4,4'-bis(2-iodoethynyl)-1,1'-biphenyl, 1,4-bis(2-iodoethynyl)benzene and 1,2-bis[4-(iodoethynyl)phenyl]-1,2-diphenylethene in dimethyformamide. The resulting polymers contained 1,3-diyne moieties and alternate aromatic rings. Poly(1,2-bis[4-(iodoethynyl) phenyl]-1,2-diphenylethene) was soluble in commonly used organic solvents, whereas the other two were insoluble owing to their rigid structures. Up to 352 °C, the polymers exhibited only 5% loss in weight. By spin coating the solution of poly(1,2-bis[4-(iodoethynyl) phenyl]-1,2-diphenylethene), a homogeneous film could be readily obtained, which possessed high refractive indices ranging from 1.7747 to 2.1115 in a wavelength range of 400 to 900 nm. Recently, Jiang *et al.* [43] also reported Fe(0)-mediated living radical polymerization to synthesize poly(glycidyl methacrylate) (PGMA) with post modification using PhSeZnCl. The resulting materials showed high refractive indices, which could be tuned by altering the doping concentration of selenide. Several other recent studies on halogen containing optical polymers have also reported refractive indices in the range of 1.7 to 2.0 and low birefringence [44-46].

4.3.4 Silicon Containing HRIPs

Heavier main group elements like tin, germanium, silicon, etc., have been recently reported for improving the refractive indices of polymers. Slow reactions between a multi-functional thiol and a main group vinyl or allyl compound can lead to polymers with highly polarizable main group elements [47-52]. Especially, silicon based HRIPs have high stability, which makes them suitable for organic light emitting diodes (OLEDs), among other uses [49]. Several recent studies have confirmed the application of silicon enriched polymers for advanced applications. For instance, using launch conditioning and refractive index engineering, low-loss and high-bandwidth multimode polymer waveguide crossings and bends with silicon have been studied [50]. In another study, highly transparent as well as highly tunable refractive index polymers were prepared using titania-polymer composites [53]. The refractive index could be tuned between 1.5 and 3.1 by changing the titania concentration. The composites were used to fabricate antireflective layers of high-resistivity silicon. By using hot embossing method, a graded index structure was fabricated utilizing the thermoplasticity of the developed composites.

Using high energy grinding method, uniformly dispersed TiO_2 nanoparticles (NPs) in silicone was fabricated for LED encapsulation [54]. At 550 nm, high refractive index of 1.63 was obtained for the silicone/TiO_2 hybrid, which was significantly higher than the refractive index of pure silicone (1.5). Thermal conductivity and barrier properties of silicon were also improved on incorporating TiO_2. Compared to pure silicone, the material exhibited better stability and 7.3% increase in the light output.

Kim *et al.* [55] reported zirconium-phenyl siloxane hybrid material (ZPH), fabricated via hydrosilylation-curing of sol-gel derived multifunctional (vinyl- and hydride-functions) siloxane resins containing phenyl-groups and Zr–O–Si heterometallic phase (Figure 4.3). The refractive index of the hybrid was observed to be 1.58. Compared with commercial LED encapsulants, high thermal stability without any yellowing for 1008 h at 180 °C and high optical transparency of ~88% at 450 nm were observed.

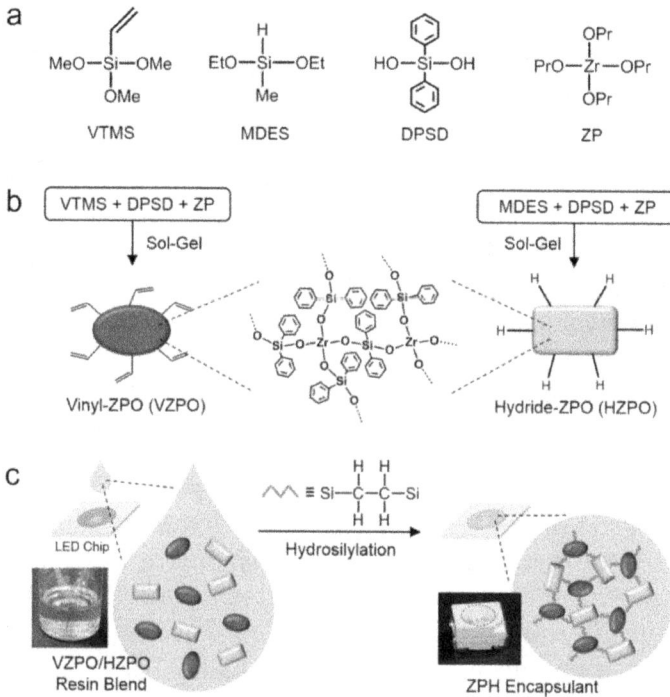

Figure 4.3 Schematic of the hybrid synthesis. Reproduced from Reference 55 with permission from American Chemical Society.

Visibly transparent porous silicon dioxide ($PSiO_2$) and $PSiO_2/TiO_2$ optical elements were fabricated by Ocier *et al.* [56]. The materials were synthesized either by thermal oxidation or a combination of atomic layer deposition and thermal oxidation infilling processes. Several other silicone based polymers with high refractive index as well as superior thermal efficiency and transparency have also been reported [57,58].

4.3.5 Other Nanocomposite HRIPs

The refractive indices of polymers can be improved by incorporating metal ions in the matrix. By using organometallic compounds, where the metal moieties are built into the polymer chains, the challenges due to proccessability and solubility can be avoided [59,60].

Poly(vinyl alcohol) (PVA) nanocomposites were generated using different weight fractions of dysprosium oxide doped with sodium ($Na_2Dy_2O_4$) [61]. On incorporation of the filler, the hydrophilic nature of the polymer became hydrophobic. Also, the electronic band gap shifted from 4.25 eV to 3.55 eV. In relation with Lorenz-Lorentz effective medium theory, the refractive index was also observed to be improved by the addition of the filler. The developed materials exhibited high potential for optical and micro-optic wave guiding applications, due to the characteristics such as high Abbe number, transparency in the visible light region and superior refractive index of 1.9 (observed for 4 wt% $Na_2Dy_2O_4$/PVA.

Block copolymers can be used for the fabrication of photonic crystals by self-assembly. These photonic crystals are useful for fast and simple device production. However, small difference in the refractive indices of the phase segregated domains limits the use of these materials for device fabrication. To overcome such issues, the optical response of the self-assembled (polynorbornene-graft-poly(tert-butyl acrylate))-block-(polynorbornene-graft-poly(ethylene oxide)) (PtBA-b-PEO) brush block copolymers (BBCPs) was tuned using gallic acid coated zirconium oxide (ZrO_2) nanoparticles by Song *et al.* [62]. The refractive index could be boosted from 1.45 to 1.70 by incorporating 70 wt% (42 vol%) ZrO_2 nanoparticles in the domain of the PEO brushes. This was made possible by the robust H-bond contacts between the ligands on ZrO_2 and PEO brushes of the BBCPs. Compared to the non-modified BBCPs, the photonic nanocomposites exhibited an enhanced reflection of around 250% at 398 nm with domain spacing of 137 nm. Another benefit of the developed nanocomposites was the reduction in the coating thickness because of the larger reflectivity due to the significant difference in the refractive indices of the domains.

For optoelectronic applications, *in-situ* polymerization of a precursor containing functional monomers and surface modified anatase TiO_2 nanoparticles was used for developing high refractive index transparent nanocomposites [63]. Due to the liquid state of the monomers, environment friendly solventless precursors could be synthesized. The TiO_2 phase contributed to enhanced thermal and mechanical properties as well as refractive index. The transparency was also observed to increase with dispersity and TiO_2 content. High dispersity and transparency with the same amount of TiO_2 loading could be attributed to a small extent of Rayleigh scattering. From 500 to 800 nm, TiO_2/poly(4-vinyl benzyl alcohol) with an 18 vol% (60 wt%) of

TiO_2 exhibited a transparency greater than 85% and a refractive index of 1.73. By substituting the aliphatic acetic acid functionalized TiO_2 with aromatic phenyl acetic acid treated TiO_2, the refractive index of the nanocomposite could be enhanced further to 1.77.

Maeda *et al.* [64] reported *in-situ* polymerization of methyl methacrylate in the presence of TiO_2 nanoparticles bearing PMMA chains grown using surface-initiated atom transfer radical polymerization. The surface modification of TiO_2 nanoparticles helped in preventing their aggregation during surface-initiated controlled polymerization of MMA. The transparency of the developed hybrids was observed to depend on the polymerization period (hence chain length) of the modified PMMA on the surface of the nanoparticles (Figure 4.4). With 6.3 vol% of TiO_2, the refractive index could be enhanced from 1.492 for pristine PMMA to up to 1.566 for the nanocomposites. Other metal based high refractive index polymer composites systems with zirconium oxide nanoparticles [65], silver nanoparticles [66], fullerenes [67], zinc bis(allyldithiocarbamate) [68], etc., have also been reported to be promising candidates for optical applications.

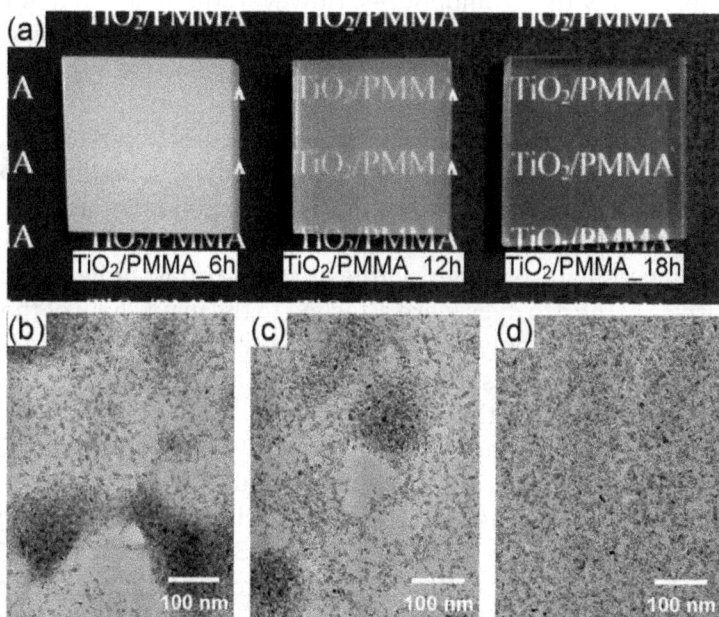

Figure 4.4 (a-c) Images showing the dependence of polymerization period (hence chain length) of PMMA on the on transparency, (b-d) represent corresponding transmission electron microscopy images. Reproduced from Reference 64 with permission from American Chemical Society.

4.4 Non-linear Optical Polymers

Non-linear optical effects refer to the changes in the amplitude frequency, phase and other propagation characteristics because of the interaction of the dielectric media in a material with the incident electromagnetic field. The displacement and charge variation of the associated atoms, owing to the interaction of the optical field with the organic molecules in the material when a beam of light propagates through the material, causes the non-linearity in the optical properties of the material [69-73]. The absence of centrosymmetry is an essential factor for developing non-linear optical effects in any material. It is applicable at both macroscopic and molecular level [73-78]. The following section describe a few non-linear optical polymers.

4.4.1 Poly(alkyl vinyl ether)s

Very few poly(vinyl ether)s with non-linear optical properties have been reported. For inducing non-linearity, these polymers are usually functionalized with chromophores. However, the relaxation of chromophores in the polymer matrix at room temperature as well as low glass transition temperature of these polymers limit their use in practical applications.

Campbell *et al.* [79] synthesized two vinyl ether polymers consisting of 4-amino-4-nitroazobenzene pendant dyes. Length of spacer from the main chain was different for the two polymers. The synthesis involved the polymerization of 2-chloroethylvinyl ether with HI/I_2. Using Williamson ether synthesis route, propyl or hexyl alcohol derivatives of the azo dyes were attached to the main chain of the polymer. By cooling the polymers from the isotropic melt phase, a liquid crystal behavior was observed. With higher azo dye loading, the material also showed high thermal stability.

In another study, reactions of 2-iodoethyl vinyl ether with 3',5'-dimethoxy-4'-hydroxy-4-dinitrostilbene and 3',5'-dimethoxy-4'-hydroxy-2,4-dinitrostilbene led to the synthesis of 3',5'-Dimethoxy-4'-(2-vinyloxyethoxy)-4-nitrostilbene   and   3',5'-dimethoxy-4'-(2-vinyloxyethoxy)-2,4-dinitrostilbene respectively [80]. Cationic initiators were used to polymerize the monomers to obtain non-linear optical phores. The resultant polymers had good thermal stability and were soluble in dimethylformamide and dimethyl sulfoxide. The authors also reported the condensation reactions of methyl cyanoacetate and malononitrile with 4-(2'-vinyl oxy ethoxy) isophthaldehyde to generate 4-di-(2'-carbomethoxy-2'-cyano vinyl)-1-(2'-vinyl oxy ethoxy) benzene and 4-di-(2',2'-dicyanovinyl)-1-(2'-vinyloxyethoxy) benzene [81]. The two monomers were polymerized with boron trifluoride etherate to obtain poly(vinyl ether)s containing oxybenzylidenecyanoacetate and oxybenzylidenemalononitrile groups. The functional groups on the main chain acted as chromophores capable of exhibiting non-linear optical properties. The glass transition temperature of the polymers was in the range of range 73-87 °C, and the polymers demonstrated good thermal stability up to 300 °C. The polymers were also observed to be soluble in common solvents like dimethyl sulforide and acetone.

Park *et al.* [82] reported the preparation of vinyl-addition poly(norbornene) copolymers, which were functionalized with non-linear chromophore side groups using (η^6-toluene)Ni(C$_6$F$_5$)$_2$. The Ni complex employed for the polymerization exhibited tolerance towards different functional groups in the non-linear optical chromophores. Compared with the methacrylate polymers, vinyl-addition copolymer of hexylnorbornene and a norbornene-functionalized Disperse Red 1 chromophore demonstrated faster polar order relaxation and poling of the electric field.

4.4.2 Polystyrenes (PS)

Glass transition temperature of polystyrene based non-linear optical chromophores is higher as compared to other polymers, which makes their device integration much easier for non-linear optical applications. The non-linearity of these polymers can be further improved by using flexible spacers, however, these spacers may impact the glass transition temperature of the polymers. Thus, an effective and optimal spacer is needed to improve the non-linearity of polystyrene based optical polymers [83,84].

Sulfonated PS matrices incorporated with CdS nanoparticles were studied for their optical properties by Du *et al.* [83]. The –SO$_3^-$ groups acted as coordination sites for *in-situ* growth of CdS particles. The concentration of the Cd^{2+} feed ions and sulfonate content of PS affected the size and density of the nanoparticles. A confined medium for particle growth with uniformity was provided by 9.9 mol% of sulfonate content which gave rise to an ionic clustering within the polymer matrix. The non-linear refractive index of the composite was observed to vary with the input irradiance (Figure 4.5), thus, indicating not just third-order but possible higher order non-linearity.

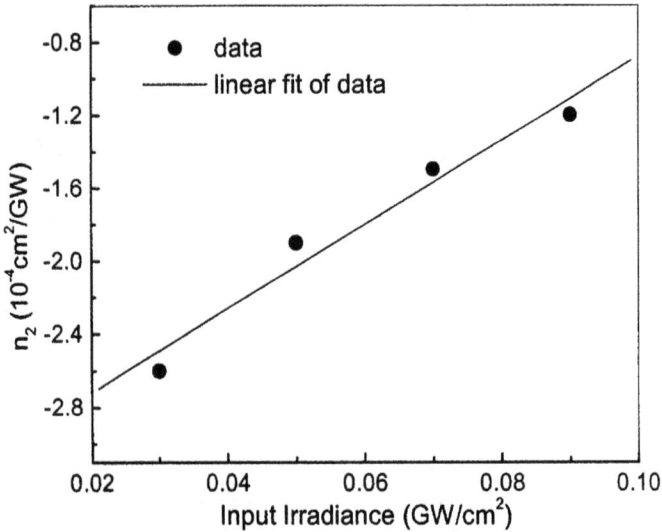

Figure 4.5 The non-linear refractive index of CdS-polystyrene composites as a function of the input irradiance. Reproduced from Reference 83 with permission from American Chemical Society.

In another study, Ag or Au nanoparticles containing polymer films were generated by reducing organometallic compounds in the polymer matrix [85]. The resulting films were observed to exhibit third order non-linear optical susceptibility. Venturini *et al.* [86] reported optical non-linearity and optical limiting behavior of pure $C_{60}-(PS_x)_y$ star polymer and polystyrene matrix incorporated with $C_{60}-(PS_x)_y$. Two, four and six polystyrene arms of controllable length were grafted to the fullerene cage, which avoided the aggregation of C_{60} molecules. The non-linear optical properties of the films were affected by the intermolecular interactions caused by the high concentration of fullerene molecules. Other non-linear polystyrene based optical systems for various device applications have also been reported [87].

4.4.3 Polymethacrylates

High quality thin films can be effortlessly achieved using methacrylate side-chain polymers. In addition, these polymers exhibit high stability towards non-linear optical applications at room temperature. At higher temperatures too, their long term stability can be improved by choosing appropriate polymerization techniques as well as fillers.

Non-linear optically active methacrylates with azo benzene side groups and bulky comonomer 1-adamantyl methacrylate were used to synthesize optically active polymethacrylates [88]. The glass transition temperatures of the polymers were found to be in the range of 160 to 190 °C. The polymers could be spin coated for obtaining thin films due to solubility in common organic solvents. At 633 nm, after electropoling with 120 V/μm electric field, an electro-optic coefficient with a value of 25 pm/V was observed. Due to the high thermal stability, glass transition temperature and mechanical stability, the polymers exhibited high potential of practical integration into real life non-linear optical devices. Nikonorova *et al.* [89] used dielectric spectroscopy to study the molecular mobility of methacrylic copolymers with different concentrations of chromophore groups. Three relaxation processes were observed in the frequency range 10^{-1}–$5 \cdot 10^{6}$ Hz at temperatures between −100 and 220 °C. The mobility of azo-chromophore groups, which is responsible for the non-linear optical properties, was observed to begin at a temperature 30 °C higher than that for α process.

In another study, polymer/metal nanohybrid systems composed of well-defined diblock copolymers and metallic palladium nanoparticles were characterized for their non-linear optical properties [90]. The copolymers were synthesized by reversible addition-fragmentation chain transfer (RAFT) polymerization. The block copolymers were based on poly[2-(N-carbazolyl) ethyl methacrylate]-block-poly[2-(acetoacetoxy) ethyl methacrylate] (CbzEMAx-b-AEMAy) and possessed carbazole and β-ketoester side-chain functionalities. The stabilization and complexation of palladium nanoparticles in tetrahydrofuran was enabled by β-ketoester groups present in the diblock copolymers. Using the Z-scan technique, non-linear optical behavior of the materials was studied both in thin film and solution. The effect of the surface plasmon resonance on the non-linear properties of the hybrids was also discussed in terms of a two-photon process.

Lee *et al.* [91] reported another palladium based polymer hybrid system and studied its non-linear optical properties. The materials were based on the sensitizer-bearing monomer palladium meso-phenoxy-tris(heptyl)porphyrin-ethylmethacrylate (PdmPH₃PMA), co-polymerized with an emitter-bearing monomer diphenylanthrancene methacrylate (DPAMA) and optically inert comonomer MMA. The method helped in preventing the agglomeration of the chromophores by covalent tethering of the appropriate chromophore pair to the polymeric backbone.

4.4.4 Other Non-linear Optical Polymers

The non-silicon systems made of Hydex® and Si_3N_4 were reported by Moss *et al.* [92] to be suitable for the generation of on-chip optical frequency comb and ultrafast optical pulse. In another study, upon suitable electromagnetic radiation, an extended π-conjugated ligand was observed to form in novel photochromic dithienylethene-based platinum (II) complexes (C^N^N)Pt(C≡C—DTE—C_6H_4—D) ((C^N^N) = 4, 4'-di (n-hexyl)-6-phenyl-2, 2'-bipyridine; D = H, NMe_2), which allowed the photoinduced switching of their second-order non-linear optical properties [93]. Sugita *et al.* [94] studied the non-linear optical susceptibility of Au nanorods coated with non-electrically poled non-linear optical polymer. The second harmonic conversion efficiency at surface plasmon resonance conditions was observed to be seventy times higher than off-resonance conditions. Alam *et al.* [95] also reported an approach to integrate low index non-linear materials with silicon photonics so as to achieve high non-linearity without suffering non-linear losses. In another study, Wu *et al.* [96] reported the incorporation of isolation chromophore into an "H" shaped non-linear optical polymer. Sekhosana *et al.* [97] reported the synthesis of 2,4,6-tris[3-thio-9,10,16,17,23,24-hexa(4-tertbutylphenoxy) phthalocyaninato ytterbium (III) chloride]-s-triazin and its lutetium counterpart. The materials were studied for non-linear optical behavior in solution and by embedding them in a polymer matrix as thin films. Compared to the solution, the complexes in poly(bisphenol A carbonate) demonstrated better optical properties. The optical activity due to the interaction between a chiral polymer and quantum dots was also reported by Oh *et al.* [98].

Third-order susceptibility of regioregular poly(3-hexylthiophene) in 530 to 1600 nm wavelength range in solution and thin films was studied using Z-scan technique by Szeremeta *et al.* [99]. Throughout the whole wavelength range, negative non-linear refraction was observed for the polymer. Two-photon and three-photon absorption regions were exhibited by the non-linear absorption process. In addition, a saturable absorption process was observed on approaching the linear absorption region. The effective multiphoton absorption cross-sections in the thin films were observed to be higher than solution (Figure 4.6). Wang *et al.* [100] also reported that the non-linear optical properties of carboxymethyl cellulose material were enhanced by generating a composite with graphene oxide. The developed composites exhibited potential for optical limiting applications.

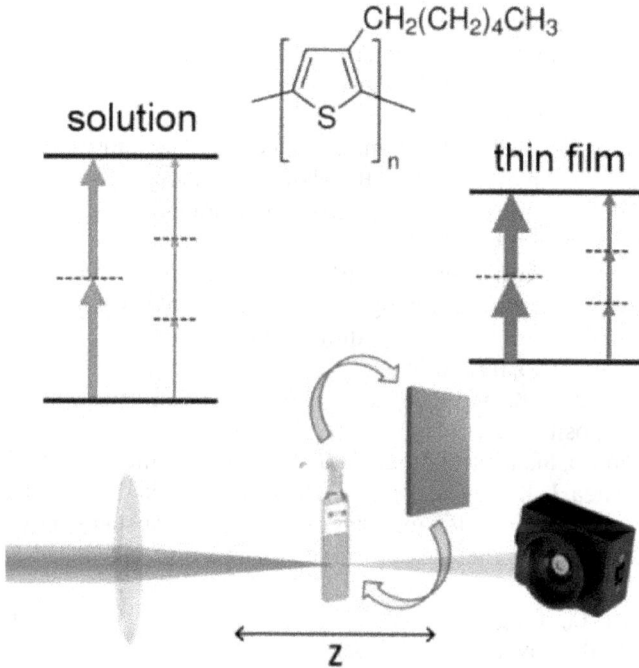

Figure 4.6 Comparison of the energy band gap in thin films and solution. Reproduced from Reference 99 with permission from American Chemical Society.

4.5 Conclusions

Polymers with optical properties have garnered a significant research attention due to efficient performance, proccessability, low cost and ease of integration in different devices. The optical properties of the polymers are dependent on various factors like chemical composition, additives, processing conditions, environment, aging, etc. Though many challenges exist with respect to the tuning of the optical properties for particular applications, however, it is envisaged that the ongoing research efforts on the optical polymers would lead to their widespread commercial uses in the near future.

References

1. Knoll, W. (2006) Optical properties of polymers. In: *Materials Sci-*

ence and Technology, Cahn, R. W., Haasen, P., and Kramer, E. J., eds., Wiley VCH, Germany, doi:10.1002/9783527603978.mst0143.

2. *Physical Properties of Polymers Handbook*, Mark, J. E. (ed.), Springer, Germany (2007).

3. Steier, W., Kalluri, S., Chen, A., Garner, S., Chuyanov, V., Ziari, M., Shi, Y., Fetterman, H., Jalali, B., Wang, W., Chen, D., and Dalton, L. R. (1995) Applications of electro-optic polymers in photonics. *MRS Proceedings*, **413**, 147.

4. Chen, J., Wieczorek, J., Eschenlohra, A., Xiao, S., and Tarasevitch, A. (2017) Ultrafast inhomogeneous magnetization dynamics analyzed by interface-sensitive nonlinear magneto-optics. *Applied Physics Letters*, **110**, 092407.

5. Thakur, V. K., and Kessler, M. R. (2015) Self-healing polymer nanocomposite materials: A review. *Polymer*, **69**, 369-383.

6. Habaza, M., Kirschbaum, M., Guernth-Marschner, C., Dardikman, G., Barnea, I., Korenstein, R., Duschl, C., and Shaked N. T. (2017) Rapid 3D refractive-index imaging of live cells in suspension without labeling using dielectrophoretic cell rotation. *Advanced Science*, **4**, 1600205.

7. Suri, G., Tyagi, M., Seshadri, G., Verma, G. L., and Khandal, R. K. (2010) Novel nanocomposite optical plastics: dispersion of titanium in polyacrylates. *Journal of Nanotechnology*, **2010**, Article ID 531284.

8. Tagaya A. (2015) Birefringence of polymer. In: *Encyclopedia of Polymeric Nanomaterials*, Kobayashi S., and Mullen K., eds., Springer, Germany.

9. Pritchard, R. (1964) The transparency of crystalline polymers. *Polymer Engineering and Science*, **4**, 66-71.

10. Boyd, R. W. J., Jones, D. C, and Webb, C. E. (2003) Nonlinear optics. In: *Handbook of Laser Technology and Applications*, Webb, C., and Jones, J., eds., Taylor & Francis, USA, pp. 161-183.

11. Gao, C., Yang, B., and Shen, J. (2000) Study on syntheses and properties of 2, 2′-mercaptoethylsulfide dimethacrylate transparent homo- and copolymer resins having high refractive index. *Journal of Applied Polymer Science*, **75**, 1474-1479.

12. Ulrich, D. R. (1990) Polymers for nonlinear optical applications. *Molecular Crystals and Liquid Crystals Incorporating Nonlinear Optics*, **189**, 3-38.

13. Born, M., and Wolf, E. (2002) *Principles of Optics*, 7th edition, Cambridge University Press, UK.

14. Tagaya, A., and Koike, Y. (2012) Compensation and control of the birefringence of polymers for photonics. *Polymer Journal*, **44**, 306-314.

15. Doi, M., and Edwards, S. F. (1986) *The Theory of Polymer Dynamics*, Oxford Science, USA.

16. Beecroft, L. L., and Ober, C. K. (1997) High refractive index polymers for optical applications. *Journal of Macromolecular Science, Part A*, **34**(4), 573-586.
17. Jha, G. S., Seshadri, G., Mohan, A., and Khandal, R. K. (2007) Development of high refractive index plastics. *e-Polymers*, **7**(1), 120.
18. Kitamura, K., Okada, K., Fujita, N., Nagasaka, Y., Ueda, M., Sekimoto, Y., and Kurata, Y. (2004) Fabrication method of double-microlens array using self-alignment technology. *Japanese Journal of Applied Physics*, **43**, 5840-5844.
19. Nakamura, T., Fujii, H., Juni, N., and Tsutsumi, N. (2006) Enhanced coupling of light from organic electroluminescent device using diffusive particle dispersed high refractive index resin substrate. *Optical Review*, **13**, 104-110.
20. Ju, Y. G., Almuneau, G., Kim, T. H., and Lee, B. W. (2006) Numerical analysis of high-index nano-composite encapsulant for light-emitting diodes. *Japanese Journal of Applied Physics*, **45**, 2546-2549.
21. Krogman, K. C., Druffel, T., and Sunkara, M. K. (2005) Anti-reflective optical coatings incorporating nanoparticles. *Nanotechnology*, **16**, S338-S343.
22. Suwa, M., Niwa, H., and Tomikawa, M. (2006) High refractive index positive tone photo-sensitive coating. *Journal of Photopolymer Science and Technology*, **19**, 275-276.
23. *Polymer Handbook*, Brandrup, J., Immergut, E. H., and Grulke, E. A. (eds.), 4th edition, John Wiley & Sons, USA (2005).
24. Hanemann, T., and Honnef, K. (2018) Optical and thermomechanical properties of doped polyfunctional acrylate copolymers. *Polymers*, **10**(3), 337.
25. Okutsu, R., Suzuki, Y., Ando, S., and Ueda, M. (2008) Poly(thioether sulfone) with high refractive index and high Abbe's number. *Macromolecules*, **41**, 6165-6168.
26. Lorentz, H. A. (1880) Ueber die Beziehung zwischen der Fortpflanzungsgeschwindigkeit des Lichtes und der Körperdichte. *Annalen der Physik*, **9**, 641-665.
27. Lorenz, L. V., (1880) Ueber die refractionsconstante. *Annalen der Physik*, **11**, 70-103
28. Liu, J. G., and Ueda, M. (2009) High refractive index polymers: Fundamental research and practical applications. *Journal of Materials Chemistry*, **19**, 8907-8919.
29. Macdonald, E. K., and Shaver, M. P. (2015) Intrinsic high refractive index polymers. *Polymer International*, **64**, 6-14.
30. *Polyimides Fundamentals and Applications*, Ghosh, M. K., and Mittal, K. L. (eds.), Marcel Dekker, USA (1996).
31. Liu, J. G., Nakamura, Y., Ogura, T., Shibasaki, Y., Ando, S., and Ueda, M. (2008) Optically transparent sulfur-containing polyimide-TiO_2 nanocomposite films with high refractive index and negative pat-

tern formation from poly(amic acid)-TiO$_2$ nanocomposite film. *Chemistry of Materials*, **20**, 273-281.

32. Griebel, J. J., Namnabat, S., Kim, E. T., Himmelhuber, R., Moronta, D. H., Chung, W. J., Simmonds, A. G., Kim, K. J., Laan, J., Nguyen, N. A., Dereniak, E. L., Mackay, M. E., Char, K., Glass, R. S., Norwood, R. A., and Pyun, J. (2014) New infrared transmitting material via inverse vulcanization of elemental sulfur to prepare high refractive index polymers. *Advanced Materials*, **26**, 3014-3018.

33. Matsuda, T., Funae, Y., Yoshida, M., Yamamoto, T., and Takaya, T. (2000) Optical material of high refractive index resin composed of sulfur-containing aliphatic and alicyclic methacrylates. *Journal of Applied Polymer Science*, **76**, 45-49.

34. Okutsu, R., Ando, S., and Ueda, M. (2008) Sulfur-containing poly(meth)acrylates with high refractive indices and high Abbe's numbers. *Chemistry of Materials*, **20**, 4017-4023.

35. Liu, J. G., Nakamura, Y., Shibasaki, Y., Ando, S., and Ueda, M. (2007) High refractive index polyimides derived from 2,7-Bis(4 aminophenylenesulfanyl)thianthrene and aromatic dianhydrides. *Macromolecules*, **40**, 4614-4620.

36. Fujita, M., Idota, N., Matsukawa, K., and Sugahara Y. (2015) Preparation of oleyl phosphate-modified TiO$_2$/poly(methyl methacrylate) hybrid thin films for investigation of their optical properties. *Journal of Nanomaterials*, **2015**, Article ID 297197.

37. Macdonald, E. K., Lacey, J. C., Ogura, I., and Shaver, M. P. (2017) Aromatic polyphosphonates as high refractive index polymers. *European Polymer Journal*, **87**, 14-23.

38. Luo, C., Zuo, J., Wang, F., Yuan, Y., Lin, F., Huang, H., and Zhao, J. (2016) High refractive index and flame retardancy of epoxy thermoset cured by tris (2-mercaptoethyl) phosphate. *Polymer Degradation and Stability*, **129**, 7-11.

39. Takahashi, S., Hotta, S., Watanabe, A., Idota, N., Matsukawa, K., and Sugahara, Y. (2017) Modification of TiO$_2$ nanoparticles with oleyl phosphate via phase transfer in the toluene–water system and application of modified nanoparticles to cyclo-olefin-polymer-based organic–inorganic hybrid films exhibiting high refractive indices. *ACS Applied Materials & Interfaces*, **9**(2), 1907-1912.

40. Tan, Z., Wu, C., Zhang, M., Lv, W., Qiua, J., and Liu, C. (2014) Phosphorus-containing polymers from tetrakis-(hydroxymethyl)phosphonium sulfate III. A new hydrolysis-resistant tris(allyloxymethyl)phosphine oxide and its thiolene reaction under ultraviolet irradiation. *RSC Advances*, **4**, 41705-41713.

41. Mehta, S., Keller, J. M., and Das, K. (2016) Nano-engineered optical properties of iodine doped poly(methyl methacrylate). *AIP Conference Proceedings*, **1731**, 080002.

42. Zhang, Y., Zhao, E., Deng, H., Lam, J. W. Y., and Tang, B. Z. (2016) De-

velopment of a transition metal-free polymerization route to functional conjugated polydiynes from a haloalkyne-based organic reaction. *Polymer Chemistry*, **7**, 2492-2500.

43. Jiang, H., Pan, X., Li, N., Zhang, Z., Zhu, J., and Zhu, X. (2017) Selenide-containing high refractive index polymer material with adjustable refractive index and Abbe's number. *Reactive and Functional Polymers*, **111**, 1-6.

44. Tapaswi, P. K., Choi, M. C., Jeong, K. M., Ando, S., and Ha, C. S. (2015) transparent aromatic polyimides derived from thiophenyl-substituted benzidines with high refractive index and small birefringence. *Macromolecules*, **48**, 3462–3474.

45. Koike, K., Teng, H., Koike, Y. and Okamoto, Y. (2014) Effect of dopant structure on refractive index and glass transition temperature of polymeric fiber-optic materials. *Polymers for Advanced Technologies*, **25**, 204-210.

46. Javadi, A., Lotf, E. A., Ataei, S. M., Zakeri, M., Nasef, M. M., Ahmad, A., and Ripin, A. (2015) High refractive index materials: A structural property comparison of sulfide- and sulfoxide-containing polyamides. *Journal of Polymer Science, Part A: Polymer Chemistry*, **53**, 2867-2877.

47. Stiegman, A. (2011) High Refractive Index Polymers, patent US2011/0054136A1.

48. Bhagat, S. D., Chatterjee, J., Chen, B., and Stiegman, A. E. (2012) High refractive index polymers based on thiol-ene cross-linking using polarizable inorganic/organic monomers. *Macromolecules*, **45**, 1174-1181.

49. Suzuki, Y., Higashihara, T., Ando, S., and Ueda, M. (2010) Synthesis of amorphous copoly (thioether sulfone) s with high refractive indices and high Abbe numbers. *European Polymer Journal*, **46**, 34-41.

50. Chen, J., Bamiedakis, N., Vasilev, P., Penty, R., and White, I. (2016) Low-loss and High-bandwidth Multimode Polymer Waveguide Components using Refractive Index Engineering, 2016 Conference on Lasers and Electro-Optics (CLEO), USA.

51. Wei, Q., Pötzsch, R., Liu, X., Komber, H., Kiriy, A., Voit, B., Will, P. A., Lenk, S., and Reineke, S. (2016) Hyperbranched polymers with high transparency and inherent high refractive index for application in organic light-emitting diodes. *Advanced Functional Materials*, **26**, 2545-2553.

52. Wei, Q., Potzsch, R., Komber, H., Pospiech, D., and Voit, B. (2014) High refractive index hyperbranched polymers with different naphthalene contents prepared through thiol-yne click reaction using disubstituted asymmetric bulky alkynes. *Polymer*, **55**, 5600-5607.

53. Wang, X., Li, Y., Cai, B., and Zhu, Y. (2015) High refractive index composite for broadband antireflection in terahertz frequency range. *Applied Physics Letters*, **106**, 231107.

54. Huang, J. H., Li, C. P., Jian, C. W. C., Lee, K. C., and Huang, J. H. (2015) Preparation and characterization of high refractive index silicone/TiO2 nanocomposites for LED encapsulants. *Journal of the Taiwan Institute of Chemical Engineers*, **46**, 168-175.
55. Kim, Y. H., Bae, J. Y., Jin, J., and Bae, B. S. (2014) Sol-gel derived transparent zirconium-phenyl siloxane hybrid for robust high refractive index LED encapsulant. *ACS Applied Materials & Interfaces*, **6**, 3115-3121.
56. Ocier, C. R., Krueger, N. A., Zhou, W., and Braun, P. V. (2017) Tunable visibly transparent optics derived from porous silicon. *ACS Photonics*, **4**, 909-914.
57. Choi, M., Leem, J. W., and Yu, J. S. (2015) Antireflective gradient-refractive-index material-distributed microstructures with high haze and superhydrophilicity for silicon-based optoelectronic applications. *RSC Advances*, **5**, 25616-25624.
58. Tan, H., Furlan, A., Li, W., Arapov, K., Santbergen, R., Wienk, M. M., Zeman, M., Smets, A. H. M., and Janssen, R. A. J. (2016) Highly efficient hybrid polymer and amorphous silicon multijunction solar cells with effective optical management. *Advanced Materials*, **28**, 2170-2177.
59. Paquet, C., Cyr, P. W., Kumacheva, E., and Manners, I. (2004) Polyferrocenes: metallopolymers with tunable and high refractive indices. *Chemical Communications*, **2**, 234-235.
60. Haubler, M., Lam, J. W. Y., Qin, A., Tse, K. K. C., Li, M. K. S., Liu, J., Jim, C. K. W., Gao, P., and Tang, B. Z. (2007) Metallized hyperbranched polydiyne: a photonic material with a large refractive index tunability and a spin-coatable catalyst for facile fabrication of carbon nanotubes. *Chemical Communications*, **25**, 2584-2586.
61. Shilpa, K. N., Nithin, K. S., Sachhidananda, S., Madhukar, B. S., and Siddaramaiah, (2017) Visibly transparent PVA/sodium doped dysprosia (Na2Dy2O4) nano composite films, with high refractive index: An optical study. *Journal of Alloys and Compounds*, **694**, 884-891.
62. Song, D. P., Li, C., Li, W., and Watkins, J. J. (2016) Block copolymer nanocomposites with high refractive index contrast for one-step photonics. *ACS Nano*, **10**, 1216-1223.
63. Tsai, C. M., Hsu, S. H., Ho, C., Tu, Y. C., Tsai, H. C., Wang, C. A., and Su, W. F. (2014) High refractive index transparent nanocomposites prepared by in situ polymerization. *Journal of Materials Chemistry C*, **2**, 2251-2258.
64. Maeda, S., Fujita, M., Idota, N., Matsukawa, K., and Sugahar, Y. (2016) Preparation of transparent bulk TiO2/PMMA hybrids with improved refractive indices via an in situ polymerization process using TiO2 nanoparticles bearing PMMA chains grown by surface-initiated atom transfer radical polymerization. *ACS Applied Materials & Interfaces*, **8**, 34762-34769.

65. Liu, C., Hajagos, T. J., Chen, D., Chen, Y., Kishpaugh, D., and Pe, Q. (2016) Efficient one-pot synthesis of colloidal zirconium oxide nanoparticles for high-refractive-index nanocomposites. *ACS Applied Materials & Interfaces*, **8**, 4795-4802.

66. Kedawat, G., Gupta, B. K., Kumar, P., Dwivedi, J., Kumar, A., Agrawal, N. K., Kumar, S. S., and Vijay, Y. K. (2014) Fabrication of a flexible UV band-pass filter using surface plasmon metal-polymer nanocomposite films for promising laser applications. *ACS Applied Materials & Interfaces*, **6**, 8407-8414.

67. Chen, S., Chen, D., Lu, M., Zhang, X., Li, H., Zhang, X., Yang, X., Li, X., Tu, Y., and Li, C. Y. (2015) Incorporating pendent fullerenes with high refractive index backbones: a conjunction effect method for high refractive index polymers. *Macromolecules*, **48**, 8480-8488.

68. Nagayama, S., and Ochiai, B. (2016) Zinc bis (allyldithiocarbamate) for highly refractive and flexible materials via the thiol-ene reaction. *Polymer Journal*, **48**, 1059-1064.

69. Attard, G. S., and Williams, G. (1986) The effect of cooling rate on the alignment of a liquid crystalline polymer as studied by dielectric relaxation spectroscopy. *Polymer Communications*, **27**, 66-68.

70. Barry, P. L., Ravaux, G., Dubois, J. C., Parneix, J. P., Njeumo, R., Legrand, C., and Levelut, A. M. (1987) Some New Side-Chain Liquid Crystalline Polymers for Non Linear Optics. *Proc. SPIE 0682, Molecular and Polymeric Optoelectronic Materials*, doi: 10.1117/12.939638.

71. Griffin, A. C., Bhatti, A. M., and Hung, R. S. L. (1987) Synthesis Of Sidechain Liquid Crystal Polymers for Nonlinear Optics. *Proc. SPIE 0682, Molecular and Polymeric Optoelectronic Materials*, doi: 10.1117/12.939639.

72. Ringsdorf, H., and Schmidt, H.-W. (1984) Electro-optical effects of azo dye containing liquid crystalline copolymers. *Macromolecular Chemistry and Physics*, **185**(7), 1327-1334.

73. Esselin, P., Le Barny, P., Robin, P., Broussoux, D., Dubois, J. C., Raffy, J., Pocholle, J. P. (1988) Second Harmonic Generation In Amorphous Polymers. *Proc. SPIE 0971, Nonlinear Optical Properties of Organic Materials*, doi: 10.1117/12.948223.

74. Boshard, Ch., Sutter, K., Pretre, Ph., Hulliger, J., Florsheimer, M., Kaatz, P., and Gunter, P. (1995) *Organic Nonlinear Optical Materials*, Gordon and Breach Publishers, Switzerland.

75. Baldwin, G. C. (1969) *An Introduction to Nonlinear Optics*, Plenum Press, USA.

76. Franken, P. A., Hill, A. E., Peters, C. W., and Weinreich, G. (1961) Generation of optical harmonics. *Physical Review Letters*, **7**, 118-119.

77. Williams, D. J. (1984) Organic polymeric and non-polymeric materials with large optical nonlinearities. *Angewandte Chemie International Edition*, **23**, 690-703.

78. *Nonlinear Optical and Electro-Active Polymers*, Prasad, P. N., and Ulrich, D. R. (eds.) Plenum Press, USA (1988).

79. Campbell, D., Dix, L. R., and Rostron, P. (1993) Synthesis of poly vinyl ethers with pendant non-linear optical azo dyes. *European Polymer Journal*, **29**, 249-253.

80. Lee, J. Y. (1995) Preparation of poly(alkylvinylether)s for nonlinear optical applications *Polymer Bulletin*, **35**, 33-40.

81. Lee, J. Y., Lee, W. J., Rhee, B. K., and Min, H. S. (2004) Synthesis of novel poly(vinyl ether)s containing the oxybenzylidenemalononitrile group as a nonlinear optical chromophore, and their electro-optic properties. *Polymer International*, **53**, 169-175.

82. Park, K. H., Twieg, R. J., Ravikiran, R., Rhodes, L. F., Shick, R. A., Yankelevich, D., and Knoesen, A. (2004) Synthesis and nonlinear-optical properties of vinyl-addition poly(norbornene)s. *Macromolecules*, **37**, 5163-5178.

83. Du, H., Xu, G. Q., and Chin, W. S. (2002) Synthesis, characterization, and nonlinear optical properties of hybridized CdS-Polystyrene nanocomposites. *Chemistry of Materials*, **14**, 4473-4479.

84. Eckenrode, H. M., and Dai, H. L. (2004) Nonlinear optical probe of biopolymer adsorption on colloidal particle surface: Poly-l-lysine on polystyrene sulfate microspheres. *Langmuir*, **20**, 9202-9209.

85. Ogawa, S., Hayashi, Y., Kobayashi, N., Tokizaki, T., and Nakamura, A. (1994) Novel preparation method of metal particles dispersed in polymer films and their third-order optical nonlinearities. *Japanese Journal of Applied Physics*, **33**, L331.

86. Venturini, J., Koudoumas, E., Couris, S., Janot, J. M., Seta, P., Mathis, C., and Leach, S. (2002) Optical limiting and nonlinear optical absorption properties of C60–polystyrene star polymer films: C60 concentration dependence. *Journal of Materials Chemistry*, **12**, 2071-2076.

87. Man, W., Fardad, S., Zhang, Z., Prakash, J., Lau, M., Zhang, P., Heinrich, M., Christodoulides, D. N., and Chen, Z. (2013) Optical nonlinearities and enhanced light transmission in soft-matter systems with tunable polarizabilities. *Physical Review Letters*, **111**, 218302.

88. Eckl, M., Muller, H., Strohriegl, P., Beckmann, S., Etzbach, K. H., Eich, M., and Vydra, J. (1995) Non-linear optically active polymethacrylates with high glass transition temperatures. *Macromolecular Chemistry and Physics*, **196**, 315-325.

89. Nikonorova, N. A., Balakina, M. Y., Fominykh, O. D., Sharipova, A. V., Vakhonina, T. A., Nazmieva, G. N., Castro, R. A., and Yakimansky, A. V. (2016) Dielectric spectroscopy and molecular modeling of branched methacrylic (co)polymers containing nonlinear optical chromophores. *Materials Chemistry and Physics*, **181**, 217-226.

90. Papagiannouli, I., Demetriou, M., Christoforou, T. K., and Couris, S., (2014) Palladium-based micellar nano hybrids: preparation and

nonlinear optical response. *RSC Advances*, **4**, 8779-8788.

91. Lee, S. H., Thevenaz, D. C., Weder, C. and Simon, Y. C. (2015), Glassy poly(methacrylate) terpolymers with covalently attached emitters and sensitizers for low-power light upconversion. *Journal of Polymer Science, Part A: Polymer Chemistry*, **53**, 1629-1639.

92. Moss, D. J., Morandotti, R., Gaet, A. L., and Lipson, M. (2013) New CMOS-compatible platforms based on silicon nitride and Hydex for nonlinear optics. *Nature Photonics*, **7**, 597-607.

93. Boixe, J., Guerchais, V., Bozec, H. L., Jacquemin, D., Amar, A., Boucekkine, A., Colombo, A., Dragonetti, C., Marinotto, D., Roberto, D., Righetto, S., and Angelis, R. D. (2014) Second-order NLO switches from molecules to polymer films based on photochromic cyclometalated platinum(II) complexes. *Journal of American Chemical Society*, **136**, 5367-5375.

94. Sugita, A., Hirabayashi, T., Ono, A., and Kawata, Y. (2014) Second Harmonic Generations from Au Nanorods Coated with Nonelectrically Poled NLO Polymer. *2014 Conference on Lasers and Electro-Optics (CLEO) - Laser Science to Photonic Applications*, USA, doi: 10.1364/CLEO_QELS.2014.FTh4K.6.

95. Alam, M. Z., Sun, X., Mojahedi, M., and Aitchison, J. S. (2015) A Nonlinear Polymer Waveguide Implemented using an Augmented Low Index Guide. *2015 Conference on Lasers and Electro-Optics (CLEO)*, USA, doi: 10.1364/CLEO_QELS.2015.FW1D.2.

96. Wu, W., Ye, C., Qin, J., and Li, Z. (2013) Introduction of an isolation chromophore into an "H"-shaped NLO polymer: Enhanced NLO effect, optical transparency and stability. *ChemPlusChem*, **78**, 1523-1529.

97. Sekhosana, K. E., Amuhaya, E., and Nyokong, T. (2015) Nanosecond nonlinear optical limiting properties of new trinuclear lanthanide phthalocyanines in solution and as thin films, *Polyhedron*, **85**, 347-354.

98. Oh, H. S., He, G. S., Law, W. C., Baev, A., Jee, H., Liu, X., Urbas, A., Lee, C.W., Choi, B. L., Swihart, M. T., and Prasad, P. N. (2014), Manipulating nanoscale interactions in a polymer nanocomposite for chiral control of linear and nonlinear optical functions. *Advanced Materials*, **26**, 1607-1611.

99. Szeremeta, J., Kolkowski, R., Nyk, M., and Samoc, M. (2013) Wavelength dependence of the complex third-order nonlinear optical susceptibility of poly(3-hexylthiophene) studied by femtosecond Z-scan in solution and thin film. *The Journal of Physical Chemistry C*, **117**, 26197-26203.

100. Wang, J., Feng, M., and Zhan, H. (2014) Preparation, characterization, and nonlinear optical properties of graphene oxide-carboxymethyl cellulose composite films. *Optics and Laser Technology*, **57**, 84-89.

5

Polymeric Self-sensing Materials

5.1 Introduction

Structural materials which display high stiffness and strength are vital for applications such as civil infrastructure, machinery, aircrafts, satellites, automobiles, helicopter blades, etc. Many of these applications require lightweight structural materials so as to achieve functional efficiency, fuel economy and transportation convenience. Composites are considered as the materials of choice for enhancing the structural performance as well as efficiency in such advanced applications. However, the detection of damage in the composites is challenging as delamination as well as fiber breakage take place within the material, and the degradation is generally not evident from outside. The damage sensing in the composite materials is generally accomplished by employing traditional non-destructive evaluation (NDE) techniques, and ultrasonic inspection is regarded as the most sensitive method out of the different traditional techniques [1]. Nevertheless, it is restricted to the recognition of flaws which are precise as well as large. For the purpose of materials safety, it is of paramount importance to identify the flaws prior to their evolution into cracks of considerable size. Structural health monitoring (SHM) system recognizes different types of damage and defects induced in the structure, along with their analysis and evaluation, for improving the reliability of the engineering structures. The current interest in the SHM systems has been centered on the application of attached or embedded sensors (like phase transformation, acoustic, dynamic response, optical fiber, piezoelectric, microelectromechanical, etc.) and tagging (by the addition of magnetic, piezoelectric or electrically conducting particles into the composite materials). The embedded sensors are difficult to repair, whereas the attached sensors encounter poor durability. As the embedded sensor size becomes bigger, the degradation of the mechanical performance turns more serious.

Haleema Saleem and Vikas Mittal*, The Petroleum Institute (part of Khalifa University of Science and Technology), Abu Dhabi, UAE
*Current address: Bletchington, Wellington County, Australia
© 2019 Central West Publishing, Australia

A recent direction in the research on structural materials is associated with the advancement of multi-functional smart materials, which, in addition to excellent structural performance, also perform non-structural functions. One such non-structural function is self-sensing, which is the property through which a material senses its own conditions like temperature, damage, stress, strain, etc. Self-sensing avoids the use of connected or fixed sensors, considering that the material itself performs as a sensor [2]. Thus, the self-sensing materials result in enhanced durability, low cost, lack of mechanical property degradation and higher sensing volume. Mostly, the self-sensing materials are based on polymer matrix composites (PMCs), due to their adaptability as well as feasibility to comprise separate phases inside the polymer matrices [3]. The main benefit of using the self-sensing conducting composites is the practicability to achieve sensing and strengthening of structures concurrently. By including a conducting element, a piezoresistivity is achieved in the composite. The developments in this field have concentrated on the utilization of fillers like carbon black, carbon fibers or metal powder [4]. Nevertheless, the usage of traditional micrometer-sized fillers generally demands higher loadings for generating a percolative network, which compromises the mechanical properties of the polymer. Short as well as continuous carbon fibers (CFs), carbon particles and carbon nano-materials like carbon nanofibers (CNFs) and carbon nanotubes (CNTs) are superior nanoscale conducting elements, which generate a conducting electrical network inside the composites at a lower fraction. When the composites encounter damage or deformation, the conducting network is disturbed, thereby causing a change in the electrical resistivity. The conducting network and the resultant variation in resistivity are greatly dependent on the conducting component type, its quantity and distribution. Overall, one of the greatest benefits of self-sensing composites is the flexibility in their design.

The self-monitoring function is carried out by regulating the electrical characteristic deviations of an electrically conductive element fixed within the matrix [5-8]. Several attributes such as temperature, chemical composition, damage, strain, stress, moisture, process condition, corrosion, electro-magnetic radiation exposure, magnetic field and corrosion can be sensed using the self-sensing materials. The sensing of temperature is important due to the fact that the thermal control is beneficial for the structural preservation, hazard mitigation as well as energy conservation. Sensing of moisture is also vital as moisture influences the efficiency of PMCs. The strain sensing is important for the structural vibration control, as vibration is regarded as a strain form, and the vibration sensing together with the vibration suppression is required for achieving vibration control, thus, resulting in hazard mitigation, performance improvement and noise reduction. Self-sensing composites have the capability for sensing their own damage and deformation, and this ability makes them exceptional materials for the health inspection of civil engineering frameworks. The stress sensing in a framework is substantial for the load monitoring, structural vibration control as well as load history recording. The approach behind the sensing of stress relates to the utilization of a computable quantity for demonstrating the presence of stress. This computable quantity is generally the voltage (for piezoelectric sensor) or an electrical resistance (ER) (for resistive or piezoresistive sensor). Resistive sensor operates with the variation in resistance upon straining, whereas the piezoresistive sensor performs with the change in resistivity consequent to straining. The piezoelectric sensor functions by the generation of voltage, related to strain. The damage sensing is critical for SHM, timely repair as well as enhancement in safety. The resistance to surface (Rs) is achieved with the electrical contacts exclusively on the single side, while the resistance to volume (Rv) is attained with the electrical contacts over the volume of the composite. Rv can be analyzed in oblique, longitudinal as well as through-thickness directions. The PMC based self-sensing materials consisting of specific reinforcing continuous fibers (such as cladded glass fibers (GF)) can perform as light guides [9-12], whereas the electrical conductivity of PMCs comprising of continuous carbon fibers (c-CFs) permits the self-sensing by means of ER analysis [2,3,13,14]. Majority of smart materials like dielectric elastomers [15], piezoelectric actuators [16], ionic polymer-metal composites (IPMCs) [17], polypyrrole actuators [18] and shape memory alloys [19] display the characteristics making them suitable for self-sensing applications.

In this chapter, different sensing techniques as well as SHM technology with respect to polymeric composite materials are outlined briefly. The chapter focuses on the recent studies on the self-sensing behavior of CFs/polymer, CNTs/polymer, GF reinforced polymer and nickel (Ni) nanowires/polymer composites, along with ionic polymer and dielectric elastomer actuators, graphene aerogel elastomers, tetrapod quantum dots (tQD)/polymer composites, zinc^{II}/polyiminofluorenes system, etc. Further, various application feasibilities of polymeric self-sensing materials are also analyzed.

5.2 Self-sensing in Different Polymeric Systems

5.2.1 CFs/Polymer Structural Composites

For structural applications, PMCs usually consist of continuous fibers like polymer, carbon and glass fibers, as continuous fibers are more efficient, as compared to short fibers. CFs are considered to be superior than GFs due to their excellent stiffness and low density. CFs are also electrically conductive, thus, their composites display electrical characteristics depending on several parameters like damage and strain, thus, enabling the composites to sense themselves by means of electrical measurement.

In a study performed by Wang *et al.* [20], the utilization of the interface between the laminae as a sensor was noted to be an effective method for the sensing of moisture, damage and temperature in the c-CFs/epoxy composites. The PMCs consisting of laminae or layers of continuous fibers are mechanically fragile at the inter-laminar interface. Direct contact among the fibers of neighboring layers happens because of the matrix flow at the time of composite preparation as well as the fiber fluctuations. The existence of direct contact is demonstrated by the fact that the volume electrical resistivity of CFs/epoxy composites in through-thickness direction is limited, although epoxy matrix is considered to be electrically insulating [21]. For the purpose of temperature sensing, the interface between the laminae performed the role of either a thermocouple junction or a thermistor. The thermocouple technique needed the fibers in the contacting layers to be different, while the thermistor method did not have this requirement. The operation of thermistor was contributed by the contact electrical resistivity of the interface between the laminae, reducing reversibly with enhancing temperature, with 0.12 eV activation energy for a cross-ply configuration. It was noted that the activation energy was considerably lesser for unidirectional configuration. Distinct CFs in the contiguous laminae provided the thermocouple function. By the application of graphitic CFs, intercalated using bromine and sodium, a thermocouple sensitivity of almost 82 µV °C^{-1} was obtained. Also, it was observed that for the unidirectional as well as cross-ply junctions, the thermocouple sensitivity was similar. The electrical resistivity analysis was also employed for examining the influence of moisture on the CFs/epoxy composites. The interfacial structure significantly influences the interface contact electrical resistivity [22,23]. For the cross-ply c-CFs/epoxy composites, the moisture was noted to exhibit a reversible impact on inter-laminar interface. An enhancement in the humidity was observed to increase the resistivity reversibly. The authors also analyzed the thermal damage of the aforementioned composites in the course of thermal cycling between 18 °C and temperatures ranging from 23 °C to 200 °C. Delamination is a typical mode of damage seen in the composites, and it could be identified by an enhancement in the contact electrical resistivity of the inter-laminae interface. It was found that the contact electrical resistivity enhanced with damage. Volume resistivity has been previously employed for identifying the delamination at the time of mechanical fatigue of CFs/epoxy composites [21], however, the contact electrical resistivity is regarded as a better measure of the damage. At the time of sensing of thermal damage, concurrent temperature detection was contributed by the thermistor operation. By the application of two cross-ply layers, a two-dimensional (2D) sensor array was obtained and illustrated to be efficient for the sensing of temperature distribution. Stress detection was contributed by the contact electrical resistivity of the laminar interface, reducing consequent to the compression action acting at the interface perpendicular direction. The impact was higher as well as repeatable for the thermoplastic (nylon 6) composites, when compared to the thermoset (epoxy) composites.

In the elastic regime, the stress as well as strain remain proportional to each other. However, in the inelastic regime, the strain is not entirely revocable. In the inelastic regime, at considerably higher stress, the damage takes place. Nevertheless, the fatigue damage may develop at lower stress amplitude. For understanding the cause for damage, the information on the stress/strain conditions at the time of damage infliction and stress/strain history before the damage infliction are beneficial. In order to correlate the damage and stress/strain information, the capability to sense stress/strain and damage is essential. By using the ER analysis, the capability of the c-CFs reinforced PMCs for self-sensing the strain as well as damage under the uniaxial tension has been illustrated by Chung and Wang [2]. It was noted that the strain was in the longitudinal direction and contributed revocable variation in the resistivity. The damage involved fiber breakage, fiber-matrix debonding as well as delamination, and it generated irrevocable variations in the resistivity. Inside a lamina having tows in identical direction, the highest electrical conductivity was in the CFs direction. However, in the transverse direction in the lamina plane, the conductivity was noted to be non-zero, due to the

connection between the fibers of adjacent tows [24]. The fiber damage reduced the fiber conductivity, thereby, lowering the longitudinal
conductivity. Nevertheless, due to the brittleness of CFs, the decrease
in the conductivity because of fiber damage before the fiber fracture
was relatively small [25]. Remarkable damage in the form of matrix
damage, inter-laminar interface degradation or delamination was
demonstrated by the through-thickness resistance. The breakage of
fibers was illustrated with the enhancement in the longitudinal resistance irrevocably [26]. The presence of the fiber-matrix debonding
as well as fiber breakage at identical location induced the debonded
segments of broken fibers to be inadequate for the reinforcement,
thereby, reducing the modulus. The longitudinal as well as through-
thickness resistivities, along with their distributions, could be examined by utilizing the four-probe technique and described in terms of
the damage and strain distributions. The longitudinal resistivity distribution analysis comprised of lesser electrical contacts, when compared to the through-thickness resistivity distribution.

Flexure is considered to be more complex than the uniaxial tension for the reason that the stress/strain is non-uniform. At the time
of flexure, a sole surface is in tension, whereas the other surface is in
compression. Due to its fast response and non-destructive nature, the
ER analysis is appealing for the damage progress investigation, as for-
merly demonstrated for the tension-tension fatigue [26]. The tension-tension fatigue damage, which is related to the tensile modulus
reduction, is illustrated by an irrevocable enhancement in the volume
electrical resistivity [26]. For flexure, the capability of the ER technique for probing the tension and compression surfaces as well as the
interior concurrently in the course of loading is specifically interesting for the investigation of damage evolution. In a related study by
Wang and Chung [27], the self-sensing in c-CFs filled PMCs has been
illustrated under flexure by the analysis of direct current (DC) ERs at
the time of flexure. It was noted that the compression Rs reduced revocably upon flexure because of the strain-induced enhancement in
the current penetration degree. However, the tension Rs enhanced
revocably because of the strain-induced reduction. Further, the authors observed that the oblique resistance reduced revocably as a result of the flexural strain. The resistance of the compression/tension
surface is a superior measure of the strain when compared to the
oblique resistance, due to the fact that it is highly responsive to
smaller deflections and is determined by the deflection in an uncom-
plicated mode. The minor damage in the form of cracking at the

compression surface is represented by the change in resistance with deflection evolving into non-linear nature. It was noted that the non-linearity degree enhanced with ultimate deflection. Nevertheless, the minor damage is readily illustrated by oblique resistance. The main damage just about failure is revealed by the compression as well as tension Rs enhancing suddenly as well as irreversibly. Onset of swift enhancement took place prior for the oblique resistance and compression Rs, when compared to the tension Rs.

For monitoring the dynamic strain as well as damage simultaneously using a single technique, a measurand, which varies in value revocably at the time of reversible straining and varies irrevocably during damage, is required. The real-time self-sensing of dynamic strain as well as static/fatigue deterioration in a continuous cross-ply [0/90] CFs/polymer composite by the ER analysis was illustrated by Wang *et al.* [13]. With a cyclic or static tensile stress along the 0° direction, the ER values in this course and at right angles to the fiber sheets were examined. Consequent to the static tension to failure, it was noted that the ER in the 0° course initially reduced and subsequently enhanced (because of 0° fiber breakage), whereas the ER at right angles to the fiber sheets enhanced monotonically. The primary reduction in the value of ER in the 0° course and the enhancement of ER at right angles to the fiber sheets were due to the enhancement in the fiber alignment degree consequent to tension, even though a decrease in the fiber residual compressive stress at the time of tension generated the earlier effect [28]. The latter effect was higher for the 0° unidirectional system, when compared to the cross-ply composites, due to the fact that the 90° fibers reduced the possibility of neighboring fiber sheets contacting each other due to the enhancement in the fiber arrangement degree. At the time of cyclic tension of [0/90] cross-ply composites, it was observed that the ER in 0° direction reduced revocably, whereas the ER at right angle to the fiber sheets enhanced revocably, while the ER value varied irreversibly by a limited amount subsequent to the first cycle in both directions. In the case of [90] unidirectional composites, the ER value in the 0° direction revocably enhanced upon tension and reversibly reduced at the time of compression, owing to piezoresistivity.

The ER analysis has been employed for strain as well as damage sensing in CFs filled PMCs. The tensile strain in the direction of the composite fibers generates a reduction in the electrical resistivity in the direction of the fibers. Damage in the delamination mode generates an enhancement in the resistivity in through-thickness direction, while the fiber breakage damage brings about an enhancement in the electrical resistivity in the fiber direction. Therefore, the ER analysis permits the concurrent damage and strain sensing. The fiber damage is regarded as more severe than the matrix damage, due to the fact that the fibers are stronger as compared to the matrix and the damage involves the matrix prior to involving the fibers. Wang and Chung [21] carried out the damage sensing in c-CFs based PMCs by ER analysis, involving the examination of the through-thickness resistance. The analysis was performed at the time of fatigue testing, and the damage to the matrix was observed to begin at a third of the fatigue life. However, the damage was noted in the course of reciprocated loading on enhancing and later reducing the stress amplitude, not beyond the elastic system [14]. The aforementioned mode of loading facilitated the differentiation between the irreversible and reversible effects. The through-thickness resistance of CFs/epoxy composite laminates was noted to be a responsive measure of the damage in the matrix. It was confirmed that the longitudinal resistance was a less responsive measure of the deterioration in the matrix. However, the matrix damage between the fibers in a layer led to the longitudinal resistance to reduce irrevocably as well as constantly.

Several studies on the use of DC ER analysis have demonstrated the efficiency of the technique for sensing the deterioration imposed by impact, flexure and tension [14,26]. Wang et al. [29] studied the influence of lay-up configuration (quasi-isotropic cross-ply and unidirectional) as well as thickness (number of laminae) on the damage self-sensing feature of c-CFs/epoxy composites. It was observed that the oblique resistance was an efficient damage indicator for all thicknesses as well as lay-up configurations. In the case of thin (8-lamina) composites, the Rs of the lower surface was an adequate measure of the damage, however, it was less responsive to minor damage as compared to the oblique resistance. It was noted that the bottom, top and oblique resistances developed monotonically with increase in the impact energy, regardless of the number of laminae or lay-up configuration. For the multi-directional composites having 8 or 16 laminae, the authors observed that the bottom resistance was more damage responsive, when compared to the top resistance. In the case of 8-lamina multi-directional composites, the bottom resistance was also observed to be more damage responsive as compared to the through-thickness and oblique resistances. Further, for the unidirectional 8-lamina composite, the bottom as well as top resistances were comparatively more damage responsive, due to the damage in the pattern of the longitudinal matrix cracking. In the case of quasi-isotropic composites of 16-lamina as well as 24-lamina, the oblique and through-thickness resistances were comparatively responsive and both enhanced monotonically with increase in the impact energy.

Capability of the CFs/polymer composite laminates for self-sensing the damage [21,28] strain [13,30] and temperature [31] by means of the DC ER analysis has been previously stated in literature. In another study, Wang *et al.* [32] analyzed the self-recognizing capacity in the cylindrical form, instead of laminates. The cylinders, generated by winding filaments, are utilized for pressure vessels and different parts of lightweight frameworks. Due to the geometry difference between the cylinder and laminate forms, the electrical contact is different for both cases, and the efficiency of the self-sensing method is also dissimilar. For monitoring the damage through the ER analysis, the authors employed a four-probe technique, along with the circumferential electrical contacts on the inner or/and outer surfaces for oblique, radial and axial resistance analyses. For determining the circumferential resistance ratio (R_2/R_1), the axial electrical contacts were employed on the exterior surface. R_2/R_1 is the ratio of the damage area's circumference resistance to a remote area from the damage zone. The damage at the time of drop impact of 10 J or lower generated a reduction in the radial resistance, however, the variation in the oblique and axial resistances was negligible. Nevertheless, the greatest deterioration at the time of drop impact beyond 10 J increased the axial, oblique and radial resistances because of delamination. Circumferential resistance ratio was noted to be the most responsive, enhancing monotonically with impact energy >1.4 J.

5.2.2 CNTs/Polymer Composites

CNTs are a thin sheet of graphene coiled to form a cylinder with the two narrow ends [33]. The single walled carbon nanotubes (SWCNTs) can be regarded as the basic structural component. On the other hand, multi-walled carbon nanotubes (MWCNTs) consist of diversified co-axial cylinders of enlarging diameter around a standard axis. The idea of SHM in the polymer composites filled with electro-conductive fillers depends on the variation in ER that a percolated structure encounters when the structure is undergoing straining [34]. Due to the fact that the network ER is a function of the strain applied to the composite, the ER could be fine-tuned for self-monitoring the strain as well as matrix deterioration [35].

In the field of resistance based sensors, the application of CNTs with significant electro-chemical characteristics displays higher potential for generating smart sensor materials for the purpose of strain sensing [36]. The CNTs/polymer composites are useful as the sensing materials for different stimuli such as pressure [37], temperature [37], gases [38,39,40], chemical vapors [41] and small scale deformation. CNTs have been extensively used because of their higher surface area and larger aspect ratio. These features make them suitable candidates for chemical functionalization, along with electro-chemical and electro-mechanical sensing transduction operations inside the composite film framework. The resultant nanocomposite thin films can be subsequently applied to all structural surfaces for actuation/sensing without hindering the fundamental behavior as well as capability. The electronic characteristics of the nanotubes are regarded to be a function of their atomic structure, and the mechanical deformity could generate changes in the nanotube conductance [42]. In addition, the nanoscale size of the nanotubes permits the fabrication of miniature sensors sensitive to the mechanical surroundings, with characteristics tailored for obtaining the strain sensors having superior behavior, as compared to the metal foil based conventional strain gauges.

The concept of employing CNTs distributed in a polymer matrix as piezoresistive sensors has been established for monitoring the failure and strain procedures in the glass fiber composites (GFCs) [34]. A significant responsiveness has been illustrated when recognizing the type as well as the advancement of matrix-controlled deterioration in GF reinforced composites [43,44]. In order to establish the strain sensors at the macro-scale, both inconstantly aligned CNTs based thin films and CNTs strengthened composites [33,35,45,46] are illustrated to be responsive to the mechanical loading that results directly in resistance variation.

A polymer nanocomposite material consisting of approx. 1% SWCNTs can evolve into a self-detecting framework, provided that an electrode is applied to the surface of the composite. A further technique is to generate CNTs filled polymer thin films on the structural surface for monitoring the framework for damage. Developing materials which could self-sense and actuate through casting the CNTs in a polymer material is an advanced technique for designing smart structures. Due to the fact that the nanotubes are homogenously distributed in the framework, there is no requirement for externally mounting the strain sensors on to the structure at different locations.

Thus, the electrical transport characteristics across distinct points on the structure can be easily monitored. When the damage, such as a void or crack, is developed, the electrical resistivity in the proximity of the defect can vary, thus, leading to its detection [33]. Ability of the distributed sensing all over the framework, contrary to the strain sensing at only individual locations, enables this approach to be very appealing, as it is generally challenging to predict where the damage initiation will take place. Hence, the CNTs filled composites with non-intrusive sensing and distributed self-diagnostics ability are appropriate for SHM.

Zhang *et al.* [33] experimentally studied the self-strain sensing ability of MWCNTs/polycarbonate nanocomposites [33]. Both sinusoidal and linear dynamic strain inputs were applied to the sample, and the ER responses were examined. It was observed that the electrical response followed the strain input wave form. The sensitivity of the nanocomposites was approximately 3.5 times higher as compared to the conventional strain gauge sensor. Hence, the minimally intrusive CNTs performed as structural reinforcement additives and also contributed a self-strain sensing ability to the nanocomposites. Nevertheless, further analysis and developments are required for the practical application of such smart structures. Ku-Herrera *et al.* [47] studied the piezoresistive performance of MWCNTs/vinyl ester composites consisting of 0.3, 0.5 and 1% CNTs loaded in compression as well as tension. The variation in the ER under the tension load was observed to be positive and displayed a direct relationship with the applied strain till failure, with marginally enhanced responsiveness for reduced CNTs content. Under compression, a non-monotonic as well as non-linear piezoresistive response was observed, with ER initially reducing in the elastic system and settling at the beginning of yielding, followed by subsequent enhancement during matrix yielding. The piezoresistive behavior of the composites was observed to be responsive to the amount of CNTs for compression as compared to tension, and the determined gauge factors were greater in the compression system. It was confirmed that the piezoresistive signal was determined by the loading type, concentration of CNTs as well as elasto-plastic nature of material, and that the ER at the time of mechanical loading could permit the self-recognition of the plastic and elastic systems of the composite.

The resistivity-strain performance of the amino-functionalized MWCNTs/polyurethane (PU)-urea elastomeric nanocomposites based sensor material having significantly lower critical percolation concentration and better distribution of MWCNTs was reported by Zhang *et al.* [46]. Consequent to deformation, the composites displayed an exponential relation between strain and resistivity, independent of the MWCNTs concentration. The exponential relationship was illustrated by a corresponding enhancement of the gap width of the tunnel junction at the time of developing strain. The mechanism of charge transport was recognized as a change generated tunneling. The observational resistivity-strain reliance was explained by analyzing the tunnel junction gap-width modulation. The percolation concept confirmed that for <5% strain values, the deformation of the conductive structure of the nanotubes governed the variation in resistivity.

In another study, An *et al.* [48] generated CNTs/GFs hierarchical composite frameworks employing an electrophoretic deposition technique to integrate CNTs into the unidirectional E-glass fibers, followed by the infusion of epoxy matrix. Resultant composites displayed an ordered framework, in which the GFs were coated with CNTs with diameters approximately 10-20 nm. The aqueous dispersions of CNTs were generated employing ozonolysis as well as ultrasonication method, and the oxidized CNTs were treated with a polyethyleneimine dendrimer for enabling the cathodic electrophoretic deposition and developing the bonding with GFs. The deposition on the fabric was attained by locating the fabric before cathode and applying a direct current field, as shown in Figure 5.1. CNT-coated GF laminates displayed variations in the ER as a function of applied shear strain and allowed the self-detection of the change between the plastic and elastic load regions.

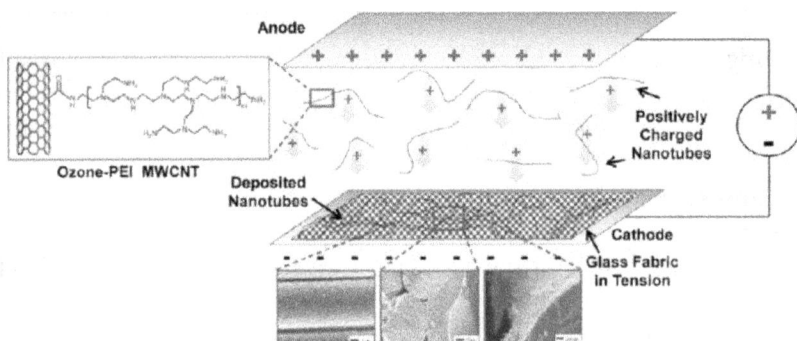

Figure 5.1 Schematic of the cathodic electrophoretic deposition on glass fabric. Reproduced from Reference 48 with permission from American Chemical Society.

Currently, the simultaneous accomplishment of greater piezoresistive sensitivity and higher strain ranges for conductive polymer nanocomposites (CPNCs) demonstrates a big challenge. In a study by Ke *et al.* [49], a facile technique for generating exceptionally piezoresistive as well as tough poly(vinylidene fluoride) (PVDF) based CPNCs was developed by adjusting the synergy between the MWCNTs and polymer matrix by employing an ionic liquid (IL) as an interface modifier/linker. Correspondingly, the IL accomplished uniform dispersion of MWCNTs in the polymer, however, induced a reduced number of CNT-CNT ohmic contacts with greater electrical contact resistance. On incorporating IL, the piezoresistive sensitivity was observed to enhance and the gauge factor changed from 7 to 60. Further, it was noted that the IL regulated CNTs-PVDF interfacial bonding, along with accomplishing remarkably enhanced sensing strain ranges and toughness of CPNCs. For understanding the cause of the remarkable dissimilarities in the piezoresistive sensitivity of the CNTs-PVDF nanocomposites, their initial electrical resistivity and CNT distribution in the polymer matrix were considered. The nanocomposites displayed greater sensitivity whenever comparatively greater electrical resistivity with loose networks existed. From Figure 5.2, it was confirmed that for both neat CNTs and IL pretreated

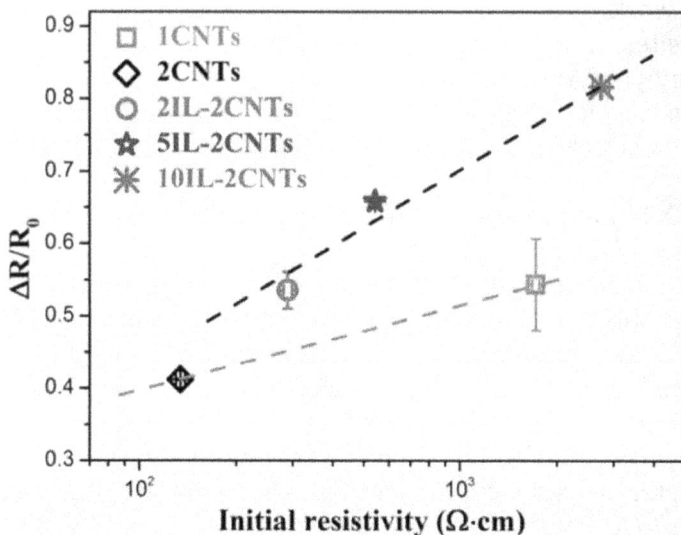

Figure 5.2 $\Delta R/R_0$ at 6% strain vs. initial electrical resistivity of PVDF nanocomposites. Reproduced from Reference 49 with permission from American Chemical Society.

CNTs, an enhancement in the sensitivity was related to the rising initial electrical resistivity. The authors also noted that the initial electrical resistivity of the nanocomposites enhanced with the IL content.

Acoustic emission (AE) is a remarkable non-destructive testing method [50], which has been employed for monitoring the fracture characteristics of the composite frameworks. During the application of tensile load to the composites, several AE signals might develop, originating from the interfacial failure, cracking of matrix as well as fiber fracture. Mostly, the AE energy generated from the fiber fracture is higher, when compared to that of debonding or matrix cracking [51]. Park *et al.* [52] carried out the self-sensing as well as interfacial analysis of the single CFs/CNTs epoxy composites having different dispersion solvents through electro-micromechanical method as well as AE under loading/consequent unloading. Advanced dispersion method was established for attaining enhanced electrical as well as mechanical properties. The apparent modulus and electrical contact resistivity of the composites were observed to associate with the dispersion solvents for CNTs. The composites employing better dispersion solvents displayed a greater apparent modulus due to the improved stress transfer effects as a result of the comparatively homogenous CNTs dispersion in the polymer and improved interfacial bonding. The good solvents, however, displayed lower thermodynamic work of adhesion for the carbon micro-fibers/CNTs-epoxy composites. Also, the damage sensing was recognized concurrently by employing AE combined with the ER analysis. Gradual enhancement in the electrical resistivity was observed with advancing fiber fracture because of the reduction in the electrical contact by CNTs.

5.2.3 CNTs Composites based on Piezoelectric Polymers

The piezoelectric materials have the exclusive characteristics of generating a differential charge on the opposite volumetric surfaces as a result of an applied mechanical deformation. The two most widely employed piezoelectric materials are PVDF copolymers and lead zirconium titanate (PZT) because of their low cost, extensive availability and ease of application. However, the non-conformable nature of PZT materials restrains their utilization on complicated structural surfaces. The conformable piezoelectric polymers such as PVDF and poly(vinylidene fluoride)/trifluoethylene P(VDF-TrFE) create minor actuation forces, however, have greater piezoelectric stress constant, which enables these materials as suitable candidates for high performance sensing applications. For the effective utilization of the piezoelectric character of the P(VDF-TrFE) and PVDF materials, an important post-fabrication process, termed as polarization or poling, is essential. Polarization is the procedure of applying an extensive electric field for inducing reorientation of the β-phase dipoles. Subsequent to the removal of the polarizing electric potential, the reoriented β-phase dipoles induce the piezoelectric effect. PVDF based polymers are currently used as inexpensive sensors in several structural vibration control applications. However, their electro-mechanical coupling coefficient has been a concern in various applications. This issue can be overcome by incorporating CNTs in the polymers, as CNTs have the ability to enhance the electro-mechanical response of PVDF based polymers. Ramaratnam and Jalili [53] illustrated that the CNTs-tuned P(VDF-TrFE) displayed superior detection ability, when compared to the pure P(VDF-TrFE) films because of the enhanced stiffness. The results confirmed that the developed voltages in the CNTs-based PVDF-TrFE composites as a result of the applied strain were greater in amplitude, as compared to the pure polymer. More precisely, it was proved that the controlling procedure accountable for enhanced detection ability was the improved Young's elastic modulus of the CNTs/polymer specimens.

In another study, Loh *et al.* [54] generated thermally evaporated as well as elevated-pressure annealed P(VDF-TrFE) nanocomposites containing different weight fractions of SWCNTs for structural monitoring applications. The films exhibited linear variation in the dielectric properties due to the applied strain. Estimated gauge factors were observed to be in the range 1.2-2.5 and were proportionate to the typical metal-foil strain gauges. It was noted that the SWCNTs/P(VDF-TrFE) composites generated similar electric capability as the commercially poled PVDF films, in spite of the fact that the SWCNTs modified films were electrode-poled with a remarkably low voltage. The capacity to produce a high electric potential under minute mechanical perturbations makes these materials suitable for self-detecting applications. As a power supply is not needed for the sensor, an information acquisition arrangement has the potential to examine numerous films concurrently in real-time by registering the output voltages generated automatically.

5.2.4 GFs Reinforced Polymer Composites

GFs reinforced polymer (GFRP) composites are widely employed for the primary as well as secondary load-bearing applications. A NDE method based on the implementation of the surface-mounted and built-in optical fiber sensors was developed in the recent past as a potential candidate for the application in damage detection and integrity monitoring of structures and materials [55,56]. The diameter discrepancy between the reinforcing and optical fibers could be overcome by employing the reinforcing GFs as the light guides (E-GFs), termed as reinforcing fiber light guides (RFLGs). A fundamental necessity for a typical light guide is that the core refractive index should be greater than the coating or cladding. The chemical constitution of the E-GFs involves almost 55% SiO_2 and numerous additives consisting of oxides of magnesium, aluminum, boron and calcium. It was illustrated that by employing RFLGs, the fiber fracture could be determined in real time by analyzing the intensity of transmitted light [57]. Owing to this, the light guides are occasionally recognized as self-sensing fibers. Practicability of the reinforcing fibers performing as the sensor system contributes several unique benefits. As an illustration, depending on the test sample length or structure to be examined, the inexpensive light sources and detectors could be utilized to analyze if the reinforcing fibers have fractured or have sustained damage. Further, the detection of damage in fiber-reinforced composites (FRCs) could be vision-based. Nevertheless, due to the lower chemical purity of E-GFs, as compared to the typical silica-based optical fibers, the transmitted light attenuation is observed to be remarkably greater.

In a study by Kister *et al.* [12], typical reinforcing E-GFs were transformed to light guides by coating the fibers with an epoxy or PU resin. The RFLGs were built-in as well as surface-mounted inside the 16-ply GFRP composites. The RFLGs were employed as sensors for detecting the deterioration generated by indentation, flexural and impact loading. Damage was confirmed favorably by checking the transmission characteristics by means of RFLGs. In spite of a remarkable enhancement in the area of delamination with developing impact energy, the transmission of light was marginally decreased, when compared to the width of the self-sensing fibers. Further, the light transmission by means of RFLGs was weakened with developing indentation load. On enhancing the flexural load, the RFLG light transmission was attenuated constantly. With respect to the surface mounted RFLGs, the impact location could be identified through bleeding light (generating from the damage area) by scanning a light beam on the surface of the composite. In another study, Brooks *et al.*

[10] surface treated typical E-GFs for enabling them to perform as light guides for shorter distances. The RFLGs were enclosed in the GFs reinforced epoxy prepregs, followed by the composite generation. The resulting composites were self-detecting, and the destruction of the fibers/interface led to the transmitted light attenuation.

Nanni *et al.* [58] analyzed the hybrid CFs-GFs reinforced polymer rods consisting of distinct CFs/GFs ratios. The generated materials displayed excellent monitoring features at the time of either pseudo-cyclic tests or static monotonic tensile tests.

5.2.5 Nickel Nanowires/Polymer Composites

The metallic nickel nanowires, generated as an extremely porous 3-dimensional (3D) lattice of inter-dependent fibers, are useful as conductive additives for polymer matrices [59,60]. Ni nanowires can be favorably combined with polymers such as acrylics, epoxies, silicone elastomers and urethanes. The Ni nanowires, as a magnetic metal, can be easily aligned under the magnetic field, further contributing several unique characteristics. Ni nanowires/polymer composites have excellent strength, rigidity as well as greater electrical conductivity at comparatively lower Ni concentrations [61]. The electrical as well as mechanical characteristics of Ni nanowires reinforced polymer composites are determined by different factors, such as degree of dispersion, intrinsic features of Ni nanowires, interfacial adhesion, aspect ratio, orientation, fiber shape and content, etc. Ni nanowires can also be employed for improving the fracture energy in the composite materials because of their effective stress transfer as well as energy adsorption mechanisms [62-64].

Park *et al.* [65] performed self-sensing as well as interfacial analysis of the nickel nanowires/polymer composites utilizing the electro-micromechanical method. The mechanical properties as well as the sensing response could be enhanced significantly with increase in the volumetric concentration of Ni nanowires. Humidity sensing analysis was carried out in a built-in humid room by setting the temperature to 30 °C. The variation in the electrical resistivity with temperature in the Ni nanowires/epoxy polymer composite with 20 vol% filler content was examined, and the electrical resistivity as well as its error range was observed to enhance with temperature. On enhancing the temperature, an enhancement in the activating mobility took place, and the stereo-regularity of Ni nanowires lattice decreased. Thus, the end-to-end lattice distance became wider, and the mean

free path and motility of electrons reduced, which accordingly increased the ER. Further, with increase in humidity, an enhancement
in the electrical resistivity was observed. With an enhancement in the
compressive loading and strain, the electrical contact resistivity was
observed to decline.

Park *et al.* [66] analyzed the self-sensing as well as actuation for
CNFs and Ni nanowires/polymer composites. The contact resistivity
and apparent modulus of CNFs/epoxy polymer composites were analyzed with respect to aspect ratio. It was noted that the CNFs/epoxy
polymer composites having lower aspect ratio displayed higher apparent modulus because of the volume increase. The Ni nan-
owires/cellulose actuators having quick frequency response and Ni
nanowires-CNFs/silicone actuators having lightweight characteristics exhibited significant advantages.

5.2.6 Ionic Polymer Actuators

The electro-mechanically active polymers (EAPs) display mechanical
deformation on electrical stimulation. Thus, these materials are beneficial for generating different types of actuators. Specifically, the
ionic electro-mechanically active polymers (IEAPs) are a major class
of EAPs, which can be employed for creating soft and small actuators
that function under low voltage (mostly 1-3 V). The IEAPs impart numerous exclusive properties to the actuators such as: simple structure that permits easy miniaturization, optionally metal-free configuration, softness, silent operation, alterable performance paths and
capability to operate in liquid surroundings. These materials also
perform as sensors for recognizing humidity [67], motion [68], curvature [69] or other ambient parameters [70].

For ionic polymer materials, the fundamental for the automatic
detection signal is usually a specific electrical specification which has
been demonstrated to associate with the mechanical performance of
the actuators. Such electrical specification could be the current developed across the IEAP electrodes because of an external manipulation
or the deformation-reliant impedance (inclusive of resistance and capacitance) [55,69] of the multilayer IEAPs or its components. The
ionic polymer metal composites (IPMCs) are a category of IEAPs in
which an ionic membrane is plated using a delicate noble metal electrode, thereby, generating a tri-layer structure [71]. The IPMCs bend
due to the minute electrical fields because of the motility of the cations present in the polymeric lattice and vice versa.

During the application of an electrical field, the transportation of the hydrated cations inside the IPMCs and the related electrostatic synergy generate bending movement of the IPMC sheets. IPMCs can be employed as sensors detecting deformation as well as motion. When externally manipulated, an electrical signal of slight amplitude is developed across the IPMC electrodes. Several theories associate the aforementioned event to a mechanically induced ion movement, which leads to the dissipation of the electrical signal [68]. For generating auto-detecting IPMCs, it is essential to determine at least a detection signal during the actuator work cycle. Due to the fact that the detection signal amplitude developed across the electrodes is modest when compared to the driving voltage, the electrical noise leads to a critical issue. The automatic detection in IPMCs can be classified into two groups, namely integrated sensor-actuator [72] and change in Rs [69]. In a study by Chen *et al.* [73], the PVDF thin films were integrated with an IPMC actuator for the purpose of sensing. In another study, Punning *et al.* [69] observed that the resistance on the surfaces of both electrodes of a sheet of IPMC material changed at the time of bending.

Nam *et al.* [74] studied the auto self-detecting capability of IPMC actuators, and the test was initiated by the Rs process variation of the electrode layers during bending. It was confirmed that the IPMC bending shift could be calculated correctly if the voltage signals on the surface of the electrode were noticed. Another simple configuration employing a half-fastened actuator as well as four-point feedback voltages was later suggested. The ability of the self-sensing method was proved from the experimental results. It was found that the suitability of the technique was restricted to the modest-bending IPMC actuators. In another study by Punning *et al.* [17], it was illustrated that the amplitude of the sensor signal was based on the curvature of bending, thus, making it feasible to recognize the actuator position. The sensor's operational convention was established on the consideration that the Rs of the metal surface electrode was determined by the bending curvature of the aforementioned sensor. From the experimental results, it was noted that the sensor signals could be effortlessly recognized and possessed a better signal to noise ratio, thus, avoiding the need for any pre-processing of the signal.

5.2.7 Dielectric Elastomer Actuators

In the past decades, numerous studies on the electro-active polymer actuators have been carried out with significant progress. Among various polymer systems, the dielectric elastomer actuators (DEA) are considered as promising candidates due to cost efficiency and large deformation. Currently, the most commonly used commercial dielectric elastomers (DE) are polyurethanes, silicones and acrylic elastomers [75]. Particularly, the capability of detecting the force or position of DEA, based on the variation of electrical characteristics like capacitance, resistance variation in accordance with the extrinsic forces, has been identified in the DEA systems based on these polymers. A unique DEA/DE sensor based self-detection technique was reported by Jung *et al.* [15] for extracting precise displacement data at the time of the actuation process without employing any extra detection equipment. The prospective self-detection performance was determined by the DEA capacitance characteristics. The DEA, along with a serial extrinsic resistor, could function as an electrical high-pass filter. The acquired voltage employing high-pass filter, essentially fabricated from DE, changed because of the variation in the overall capacitance, while the DEA was extended electro-mechanically. In order to concurrently analyze the detection as well as actuation using DEA, the authors employed a modulation approach for mixing signals, having high frequency with small amplitude signal for sensing and lower frequency signal for actuating.

Chuc *et al.* [76] developed DEA with a capability for force sensing, with no additionally added sensing equipment. The fundamental physical characteristics of the DE were experimentally analyzed, and it was observed that the impedance of the DE changed contingent upon the external forces acting on it. Based on the experimental findings, the authors proposed the self-sensing actuator. Further, a multi-stacked actuator having self-sensing ability was recognized for feasibility validation.

The inherent viscoelastic drift of the DE actuators is a significant disadvantage, and the closed-loop operation of the DE actuators is necessary for all precise uses. Rosset *et al.* [77] demonstrated the capacitive self-detection for driving a DE actuator in closed-loop, with no requirement for additional sensors. The technique was illustrated on a DE actuator tunable grating established on very high bond (VHB) acrylic as well as silicone membranes. The results confirmed that the commonly employed VHB demonstrated a time-dependent drift across the strain and capacitance of electrodes. Silicone based grating did not exhibit any drift, and the strain could be balanced by governing the device capacitance to a consistent value.

5.2.8 Graphene Aerogel Elastomers

As compared to the CNT monoliths, the graphene-based materials, particularly graphene aerogels (GA), have lower density (0.16 mg/cm^3), excellent electrical conductivity and higher compressibility (90%) [78]. The three-dimensional (3D) GA are promising candidates for use in gas sensors, electrode materials, supercapacitors and oil absorption because of their good mechanical strength, high porosity as well as electrical conductivity. Nevertheless, the actuation, control and response characteristics of GA have not been studied in detail.

Xu *et al.* [79] prepared 3D GA decorated with the Fe$_3$O$_4$ nanoparticles by self-congregation of graphene with concurrent decoration by the Fe$_3$O$_4$ nanoparticles (Fe$_3$O$_4$/GA) utilizing a modified hydro-thermal reduction method. Figure 5.3 presents the morphological analysis of the developed system. In the course of 20 compression cycles,

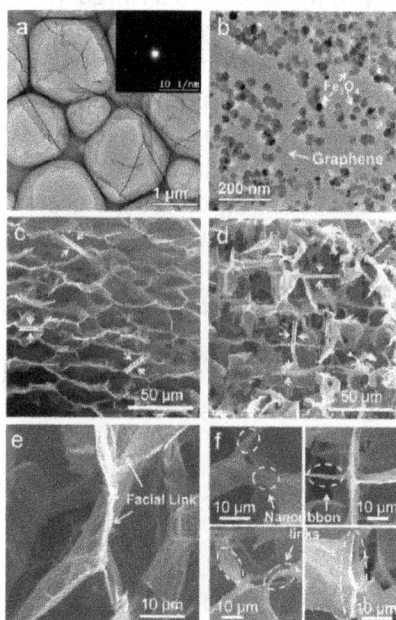

Figure 5.3 Transmission and scanning electron micrographs (TEM and SEM) images of GA and Fe$_3$O$_4$/GA: (a) TEM image of graphene oxide (GO) and the diffraction pattern of single flake (inset), (b) TEM image of Fe$_3$O$_4$ nanoparticles decorated graphene sheets, (c-d) SEM images of microporous structures of GA and Fe$_3$O$_4$/GA and (e-f) SEM images of crosslinking patterns of GA and Fe$_3$O$_4$/GA. Reproduced from Reference 79 with permission from American Chemical Society.

the resistance exhibited a synchronous as well as stable strain response. Therefore, the field-induced deformation of Fe_3O_4 nanoparticles could be noticed in real time by analyzing the changes in ER, thus, enabling the utilization of the developed material in self-detecting soft actuators as well as micro-switches. The aerogels displayed almost 52% reversible magnetic field-generated strain as well as strain-dependent ER which could be employed for monitoring the compression or stretching degree of the material. The density of the Fe_3O_4/GA material was 5.8 mg/cm^3, thus, confirming its ultra-light nature.

5.2.9 Tetrapod Quantum Dots/Polymer Composites

A visible-light nanoscale self-detecting stress probe is advantageous in a variety of engineering as well as imaging applications. Such sensor has the capability to be fixed into a range of smart structural materials without generating any degradation, and is specifically attractive for the possible sensing of impending fractures during service. Due to their unique shape as well as nanoscale size, tQDs can contribute an excellent spatial resolution of stresses when compared to other existing technologies [80]. The ability of tQDs to self-sense their nanoscale dispersion in the composites generates an exclusive optical nano-sensor which can report its own local and macroscopic characteristics, a potentially significant feature for the design of "smart" polymer nanocomposites. Raja *et al.* [81] demonstrated the tQDs based detector for the sensing of nanoscale tensile as well as compressive stresses, when fixed into smart structural block copolymer films (Figure 5.4). The authors illustrated the sensing in terms of both

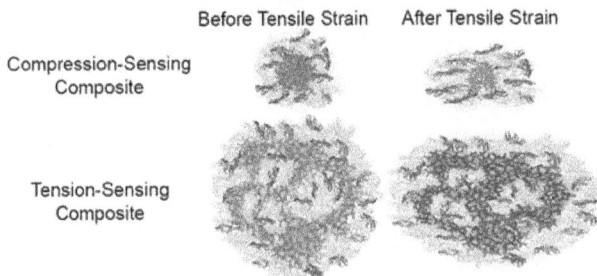

Figure 5.4 Schematic representation of the microstructures of tQDs-SEBS compression-sensing film (a) before and (b) after the application of tensile strain. Reproduced from Reference 81 with permission from American Chemical Society.

full-width at half maximum (FWHM) and photoluminescence emission-maximum. The composites were generated by mixing tQDs in chloroform with a structural block copolymer, poly(styrene-ethylene-butylene-styrene) (SEBS).

Raja *et al.* [82] also studied the cadmium selenide-cadmium sulfide (CdSe-CdS) tQDs, included into polymer matrices through electro-spinning technique, as shown in Figure 5.5, as an *in-situ* luminescent stress probe. Varying concentrations of tQDs were incorporated

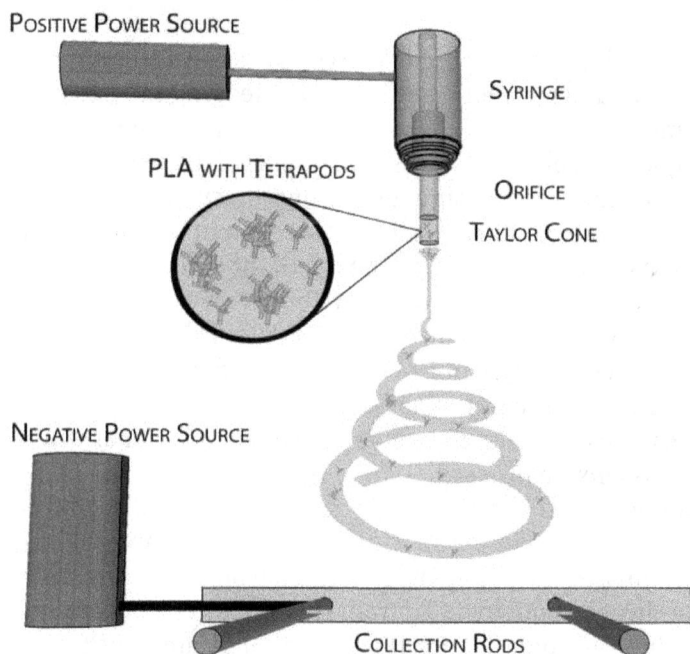

Figure 5.5 Schematic of the electro-spinning process. Reproduced from Reference 82 with permission from American Chemical Society.

in poly-l-lactic acid, generating the nanocomposites with tQDs as the nanofiller and poly-l-lactic acid as the polymer matrix. The mechanical as well as optical experiments on the nanocomposites confirmed that the tetrapod nanocrystal sensor coincided with the bulk mechanical sensor in the tensile mechanical characteristics. Certain differences between the sensing performance of the tQDs nanoscale load cell and universal testing machine macro-scale load cell were noted, which were caused by the flawed polymer-nanocrystal interface and resultant deficient stress transfer to the tQDs [83]. The particle aggregation at the time of the composite generation restricted the transfer of stress to the tQDs, which ensured the elasticity as well as recyclability of the probe. In addition, the enhancement in the concentration of the tetrapod material afforded a lower degree of change in the structural and mechanical characteristics of the polymer, however, efficiently enhanced the tQD sensor response as well as sensitivity.

5.2.10 Zinc[II]/Polyiminofluorenes System

The constitutional dynamic chemistry [84] shows adaptative performance, as its units are efficient in acknowledging the extrinsic effectors by means of constitution reorganization as a consequence of reshuffling and incorporation of components. This leads to the development on either the supramolecular or molecular level, the units being linked by means of reversible non-covalent or covalent bonding interactions [85], respectively. The resultant constitutional plasticity develops dynamic diversity and contributes access to a series of characteristics. On the molecular level, the constitutional dynamic chemistry is enforced in dynamic combinatorial chemistry (DCC) [86]. A dynamic structure might experience constitutional restructuring in response to an extrinsic effector so as to develop a signal, thus, adding up to a self-sensing process.

Giuseppone and Lehn [87] illustrated the capability of a series of oligo- as well as polyimine species. The authors analyzed the component exchange in the polymers generated from a 1:1:1 mixture of 2,7-fluorene-dicarboxaldehyde A, trans-1,4-cyclo-hexanediamine B and 2,7-diamino-fluorene C (CDL II) in ethanol at ambient temperature (Figure 5.6). A synergistic adaptative performance was observed in the developed materials, i.e., the incorporation of an extrinsic effector led to the development of the dynamic mixture towards the selection as well as augmentation of that species which permitted the signal development, thus, confirming the presence of the very effector which encouraged its generation in the first place. The observed phenomenon represents a constitutional dynamic self-detecting mechanism which enhances the scope of the functional self-sensing processes. The designed restructuring of systems is of high significance for developing active smart materials, thus, permitting the expression as well as adjustment of a specific fundamental characteristic or/and generating an adaptative reaction under the pressure of extrinsic conditions.

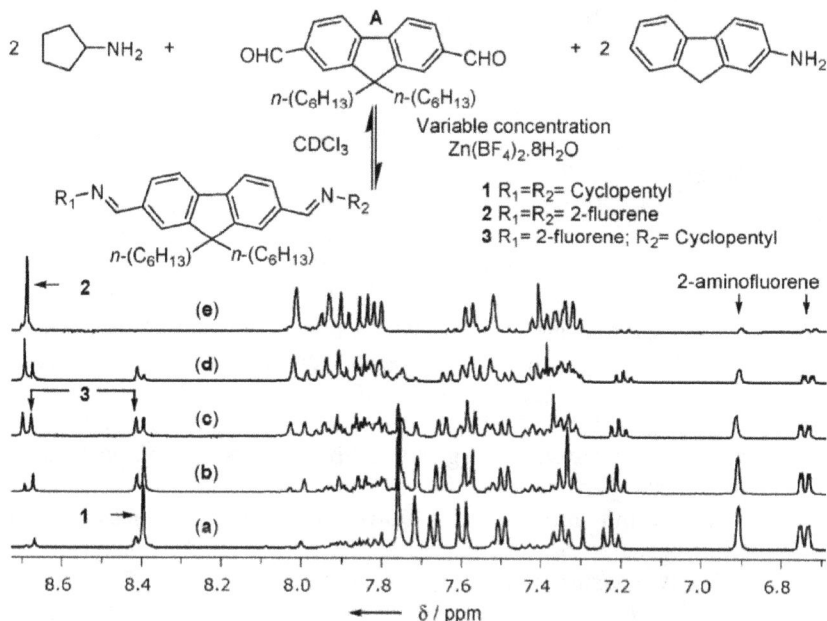

Figure 5.6 (Top) Constitutional dynamic library of iminofluorenes, CDL I, with Zn(BF₄)₂.8H₂O promoted component exchange and (bottom) ¹H NMR spectra (400 MHz) of the CDL for different Zn(BF₄)₂.8H₂O concentrations at 298 K in CDCl₃: (a) 0 M, (b) 3.10×10^{-2} M, (c) 6.20×10^{-2} M, (d) 8.68×10^{-2} M and (e) 1.24×10^{-1} M. Reproduced from Reference 87 with permission from American Chemical Society.

5.3 Summary

In this chapter, an up-to-date analysis of the different self-sensing polymeric systems and their application potential has been presented. Specifically, an overview of the different materials, sensing techniques and technology in the field of SHM with respect to the polymeric composite materials has been provided.

References

1. De Freitas, M., Silva, A., and Reis, L. (2000) Numerical evaluation of failure mechanisms on composite specimens subjected to impact loading. *Composites Part B: Engineering*, **31**(3), 199-207.
2. Chung, D. D. L., and Wang, S. (2003) Self-sensing of damage and

strain in carbon fiber polymer-matrix structural composites by electrical resistance measurement. *Polymers and Polymer Composites*, **11**(7), 515-525.

3. Kemp, R. M. J., Williamson, N. J., and Curtis, P. T. (1996) Self-sensing smart polymer composites. *IMech E Seminar, Aircraft Structures and Materials*, **10**, 47-54.

4. Flandin, L., Hiltner, A., and Baer, E. (2001) Interrelationships between electrical and mechanical properties of a carbon black-filled ethylene–octene elastomer. *Polymer*, **42**(2), 827-838.

5. Lundberg, B., and Sundqvist, B. (1986) Resistivity of a composite conducting polymer as a function of temperature, pressure, and environment: applications as a pressure and gas concentration transducer. *Journal of Applied Physics*, **60**(3), 1074-1079.

6. Prabhakaran, R. (1990) Damage assessment through electrical resistance measurement in graphite fiber-reinforced composites. *Experimental Techniques*, **14**(1), 16-20.

7. Wang, S., and Chung, D. D. L. (1998) Interlaminar interface in carbon fiber polymer-matrix composites, studied by contact electrical resistivity measurement. *Composite Interfaces*, **6**(6), 497-505.

8. Muto, N., Arai, Y., Shin, S. G., Matsubara, H., Yanagida, H., Sugita, M., and Nakatsuji, T. (2001) Hybrid composites with self-diagnosing function for preventing fatal fracture. *Composites Science and Technology*, **61**(6), 875-883.

9. Hayes, S. A., Brooks, D., Liu, T., Vickers, S., and Fernando, G. F. (1996) In-situ Self-sensing Fiber Reinforced Composites. *Proceedings of SPIE 2718, Smart Structures and Materials 1996: Smart Sensing, Processing, and Instrumentation*, USA, doi: 10.1117/12.240877.

10. Brooks, D., Hayes, S. A., Khan, N. A., Zolfaghar, K., and Fernando, G. F. (1997) Self-sensing E-glass-fiber-reinforced Composites. *Proceedings of SPIE 3042, Smart Structures and Materials 1997: Smart Sensing, Processing, and Instrumentation*, USA, doi: 10.1117/12.275729.

11. Zolfaghar, K., Khan, N., Brooks, D., Hayes, S., Liu, T., Roca, J., Gerard, F., and Bellingham, W. A. (1998) Quartz and E-glass Fiber Self-sensing Composites. *Proceedings of SPIE 3321, 1996 Symposium on Smart Materials, Structures, and MEMS*, India, doi: 10.1117/12.305565.

12. Kister, G., Ralph, B., and Fernando, G. F. (2004) Damage detection in glass fibre-reinforced plastic composites using self-sensing E-glass fibres. *Smart Materials and Structures*, **13**(5), 1166-1175.

13. Wang, X., and Chung, D. D. L. (1998) Self-monitoring of fatigue damage and dynamic strain in carbon fiber polymer-matrix composite. *Composites Part B: Engineering*, **29**(1), 63-73.

14. Wang, S., and Chung, D. D. L. (2002) Mechanical damage in carbon fiber polymer-matrix composite, studied by electrical resistance

measurement. *Composite Interfaces*, **9**(1), 51-60.

15. Jung, K., Kim, K. J., and Choi, H. R. (2008) A self-sensing dielectric elastomer actuator. *Sensors and Actuators A: Physical*, **143**(2), 343-351.

16. Anderson, E. H., and Hagood, N. W. (1994) Simultaneous piezoelectric sensing/actuation: analysis and application to controlled structures. *Journal of Sound and Vibration*, **174**(5), 617-639.

17. Punning, A., Kruusmaa, M., and Aabloo, A. (2007) A self-sensing ion conducting polymer metal composite (IPMC) actuator. *Sensors and Actuators A: Physical*, **136**(2), 656-664.

18. Otero, T. F. (2009) Soft, wet, and reactive polymers. Sensing artificial muscles and conformational energy. *Journal of Materials Chemistry*, **19**(6), 681-689.

19. Lan, C. C., Lin, C. M., and Fan, C. H. (2011) A self-sensing microgripper module with wide handling ranges. *IEEE/ASME Transactions on Mechatronics*, **16**(1), 141-150.

20. Wang, S., Kowalik, D. P., and Chung, D. D. L. (2004) Self-sensing attained in carbon-fiber–polymer-matrix structural composites by using the interlaminar interface as a sensor. *Smart Materials and Structures*, **13**(3), 570-592.

21. Wang, X., and Chung, D. D. L. (1997) Sensing delamination in a carbon fiber polymer-matrix composite during fatigue by electrical resistance measurement. *Polymer Composites*, **18**(6), 692-700.

22. Wang, S., and Chung, D. D. L. (2001) Thermal fatigue in carbon fibre polymer-matrix composites, monitored in real time by electrical resistance measurements. *Polymers and Polymer Composites*, **9**(2), 135-140.

23. Chung, D. D. L. (2001) Continuous carbon fiber polymer-matrix composites and their joints, studied by electrical measurements. *Polymer Composites*, **22**(2), 250-270.

24. Wang, X., and Chung, D. D. L. (1997) An electromechanical study of the transverse behavior of carbon fiber polymer-matrix composite. *Composite Interfaces*, **5**(3), 191-199.

25. Wang, X., and Chung, D. D. L. (1997) Electromechanical behavior of carbon fiber. *Carbon*, **35**(5), 706-709.

26. Wang, X., and Chung, D. D. L. (1999) Fiber breakage in polymer-matrix composite during static and fatigue loading, observed by electrical resistance measurement. *Journal of Materials Research*, **14**(11), 4224-4229.

27. Wang, S., and Chung, D. D. L. (2006) Self-sensing of flexural strain and damage in carbon fiber polymer-matrix composite by electrical resistance measurement. *Carbon*, **44**(13), 2739-2751.

28. Wang, X., and Chung, D. D. L. (1997) Real-time monitoring of fatigue damage and dynamic strain in carbon fiber polymer-matrix composite by electrical resistance measurement. *Smart Materials and Str-*

ctures, **6**(4), 504-508.

29. Wang, S., Chung, D. D. L., and Chung, J. H. (2005) Effects of composite lay-up configuration and thickness on the damage self-sensing behavior of carbon fiber polymer-matrix composite. *Journal of Materials Science*, **40**(3), 561-568.

30. Wang, X., and Chung, D. D. L. (1996) Continuous carbon fibre epoxy-matrix composite as a sensor of its own strain. *Smart Materials and Structures*, **5**(6), 796-800.

31. Wang, S., and Chung, D. D. L. (1999) Temperature/light sensing using carbon fiber polymer-matrix composite. *Composites Part B: Engineering*, **30**(6), 591-601.

32. Wang, S., Chung, D. D. L., and Chung, J. H. (2006) Self-sensing of damage in carbon fiber polymer-matrix composite cylinder by electrical resistance measurement. *Journal of Intelligent Material Systems and Structures*, **17**(1), 57-62.

33. Zhang, W., Suhr, J., and Koratkar, N. (2006) Carbon nanotube/polycarbonate composites as multifunctional strain sensors. *Journal of Nanoscience and Nanotechnology*, **6**(4), 960-964.

34. Fiedler, B., Gojny, F. H., Wichmann, M. H., Bauhofer, W., and Schulte, K. (2004) Can carbon nanotubes be used to sense damage in composites? *Lavoisier Annales de Chimie*, **29**(6), 81-94.

35. Kang, I., Schulz, M. J., Kim, J. H., Shanov, V., and Shi, D. (2006) A carbon nanotube strain sensor for structural health monitoring. *Smart Materials and Structures*, **15**(3), 737-748.

36. Baughman, R. H., Zakhidov, A. A., and De Heer, W. A. (2002) Carbon nanotubes - the route toward applications. *Science*, **297**(5582), 787-792.

37. Jiang, M. J., Dang, Z. M., and Xu, H. P. (2006) Significant temperature and pressure sensitivities of electrical properties in chemically modified multiwall carbon nanotube/methylvinyl silicone rubber nanocomposites. *Applied Physics Letters*, **89**(18), 182902.

38. Philip, B., Abraham, J. K., Chandrasekhar, A., and Varadan, V. K. (2003) Carbon nanotube/PMMA composite thin films for gas-sensing applications. *Smart Materials and Structures*, **12**(6), 935-939.

39. Abraham, J. K., Philip, B., Witchurch, A., Varadan, V. K., and Reddy, C. C. (2004) A compact wireless gas sensor using a carbon nanotube/PMMA thin film chemiresistor. *Smart Materials and Structures*, **13**(5), 1045-1049.

40. Zhang, B., Fu, R. W., Zhang, M. Q., Dong, X. M., Lan, P. L., and Qiu, J. S. (2005) Preparation and characterization of gas-sensitive composites from multi-walled carbon nanotubes/polystyrene. *Sensors and Actuators B: Chemical*, **109**(2), 323-328.

41. Wei, C., Dai, L., Roy, A., and Tolle, T. B. (2006) Multifunctional chemical vapor sensors of aligned carbon nanotube and polymer composites. *Journal of the American Chemical Society*, **128**(5), 1412-1413.

42. Singh, G., Rice, P., and Mahajan, R. L. (2007) Fabrication and mechanical characterization of a force sensor based on an individual carbon nanotube. *Nanotechnology*, **18**(47), 475501.

43. Thostenson, E. T., and Chou, T. W. (2006) Carbon nanotube networks: sensing of distributed strain and damage for life prediction and self-healing. *Advanced Materials*, **18**(21), 2837-2841.

44. Thostenson, E. T., and Chou, T. W. (2008) Real-time in situ sensing of damage evolution in advanced fiber composites using carbon nanotube networks. *Nanotechnology*, **19**(21), 215713.

45. Kang, I. P., Lee, J. W., Choi, G. R., Jung, J. Y., Hwang, S. H., Choi, Y. S., Yoon, K.J. and Schulz, M. J. (2006) Structural health monitoring based on electrical impedance of a carbon nanotube neuron. *Key Engineering Materials*, **321-323**, 140-145.

46. Zhang, R., Baxendale, M., and Peijs, T. (2007) Universal resistivity-strain dependence of carbon nanotube/polymer composites. *Physical Reviews B*, **76**(19), 195433.

47. Ku-Herrera, J. J., Avilés, F., and Seidel, G. D. (2013) Self-sensing of elastic strain, matrix yielding and plasticity in multiwall carbon nanotube/vinyl ester composites. *Smart Materials and Structures*, **22**(8), 085003.

48. An, Q., Rider, A. N., and Thostenson, E. T. (2013) Hierarchical composite structures prepared by electrophoretic deposition of carbon nanotubes onto glass fibers. *ACS Applied Materials and Interfaces*, **5**(6), 2022-2032.

49. Ke, K., Pötschke, P., Gao, S., and Voit, B. (2017) An ionic liquid as interface linker for tuning piezoresistive sensitivity and toughness in poly (vinylidene fluoride)/carbon nanotube composites. *ACS Applied Materials and Interfaces*, **9**(6), 5437-5446.

50. Park, J. M., Chong, E. M., Yoon, D. J., and Lee, J. H. (1998) Interfacial properties of two SiC fiber–reinforced polycarbonate composites using the fragmentation test and acoustic emission. *Polymer Composites*, **19**(6), 747-758.

51. Ma, B. T., Schadler, L. S., Laird, C., and Figueroa, J. C. (1990) Acoustic emission in single filament carbon/polycarbonate and Kevlar®/polycarbonate composites under tensile deformation. *Polymer Composites*, **11**(4), 211-216.

52. Park, J. M., Kim, P. G., Jang, J. H., Wang, Z., Kim, J. W., Lee, W. I., Park, J.G. and DeVries, K. L. (2008) Self-sensing and dispersive evaluation of single carbon fiber/carbon nanotube (CNT)-epoxy composites using electro-micromechanical technique and nondestructive acoustic emission. *Composites, Part B: Engineering*, **39**(7), 1170-1182.

53. Ramaratnam, A., and Jalili, N. (2006) Reinforcement of piezoelectric polymers with carbon nanotubes: pathway to next-generation sensors. *Journal of Intelligent Material Systems and Structures*, **17**(3),

199-208.

54. Loh, K. J., Kim, J., and Lynch, J. P. (2008) Self-sensing and power harvesting carbon nanotube-composites based on piezoelectric polymers. In: *Bridge Maintenance, Safety, Management, Health Monitoring and Informatics*, Taylor and Francis, UK, pp. 3329-3336.

55. Kruusamae, K., Punning, A., and Aabloo, A. (2012) Electrical model of a carbon-polymer composite (CPC) collision detector. *Sensors*, **12**(2), 1950-1966.

56. Wood, K., Brown, T., Rogowski, R., and Jensen, B. (2000) Fiber optic sensors for health monitoring of morphing airframes: I. Bragg grating strain and temperature sensor. *Smart Materials and Structures*, **9**(2), 163-169.

57. Kister, G., Wang, L., Ralph, B., and Fernando, G. F. (2003) Self-sensing E-glass fibres. *Optical Materials*, **21**(4), 713-727.

58. Nanni, F., Ruscito, G., Forte, G., and Gusmano, G. (2007) Design, manufacture and testing of self-sensing carbon fibre–glass fibre reinforced polymer rods. *Smart Materials and Structures*, **16**(6), 2368-2374.

59. Lin, S. W., Chang, S. C., Liu, R. S., Hu, S. F., and Jan, N. T. (2004) Fabrication and magnetic properties of nickel nanowires. *Journal of Magnetism and Magnetic Materials*, **282**, 28-31.

60. Zhang, H. Y., Gu, X., Zhang, X. H., Ye, X., and Gong, X. G. (2004) Structures and properties of Ni nanowires. *Physics Letters A*, **331**(5), 332-336.

61. Hansen, G. (2005) High aspect ratio sub-micron and nano-scale metal filaments. *SAMPE Journal*, **41**(2), 24-33.

62. Jin, C. G., Liu, W. F., Jia, C., Xiang, X. Q., Cai, W. L., Yao, L. Z., and Li, X. G. (2003) High-filling, large-area Ni nanowire arrays and the magnetic properties. *Journal of Crystal Growth*, **258**(3), 337-341.

63. Wang, S., Lee, S. I., Chung, D. D. L., and Park, J. M. (2001) Load transfer from fiber to polymer matrix, studied by measuring the apparent elastic modulus of carbon fiber embedded in epoxy. *Composite Interfaces*, **8**(6), 435-441.

64. Park, J. M., Lee, S. I., and Choi, J. H. (2005) Cure monitoring and residual stress sensing of single-carbon fiber reinforced epoxy composites using electrical resistivity measurement. *Composites Science and Technology*, **65**(3), 571-580.

65. Park, J. M., Kim, S. J., Yoon, D. J., Hansen, G., and DeVries, K. L. (2007) Self-sensing and interfacial evaluation of Ni nanowire/polymer composites using electro-micromechanical technique. *Composites Science and Technology*, **67**(10), 2121-2134.

66. Park, J.-M., Kim, P.-G., Jang, J.-H., Kim, S.-J., Yoon, D.-J., Hansen, G., and DeVries, K. L. (2007) Self-sensing of CNF and Ni Nanowire/PVDF and Cellulose Composites using Electro-micromechanical Test. *Proceedings of SPIE 6645, Nanoengineering: Fabrication, Properties,*

Optics, and Devices IV, 66451D*, USA, doi: 10.1117/12.731389.

67. Must, I., Johanson, U., Kaasik, F., Põldsalu, I., Punning, A., and Aabloo, A. (2013) Charging a supercapacitor-like laminate with ambient moisture: from a humidity sensor to an energy harvester. *Physical Chemistry Chemical Physics*, **15**(24), 9605-9614.

68. Pugal, D., Jung, K., Aabloo, A., and Kim, K. J. (2010) Ionic polymer–metal composite mechanoelectrical transduction: review and perspectives. *Polymer International,* **59**(3), 279-289.

69. Punning, A., Kruusmaa, M., and Aabloo, A. (2007) Surface resistance experiments with IPMC sensors and actuators. *Sensors and Actuators A: Physical*, **133**(1), 200-209.

70. Martinez, J. G., and Otero, T. F. (2012) Biomimetic dual sensing-actuators: theoretical description. Sensing electrolyte concentration and driving current. *The Journal of Physical Chemistry B*, **116**(30), 9223-9230.

71. Jo, C., Pugal, D., Oh, I. K., Kim, K. J., and Asaka, K. (2013) Recent advances in ionic polymer–metal composite actuators and their modeling and applications. *Progress in Polymer Science*, **38**(7), 1037-1066.

72. Bonomo, C., Brunetto, P., Fortuna, L., Giannone, P., Graziani, S., and Strazzeri, S. (2008) A tactile sensor for biomedical applications based on IPMCs. *IEEE Sensors Journal*, **8**(8), 1486-1493.

73. Chen, Z., Kwon, K. Y., and Tan, X. (2008) Integrated IPMC/PVDF sensory actuator and its validation in feedback control. *Sensors and Actuators A: Physical*, **144**(2), 231-241.

74. Nam, D. N. C., and Ahn, K. K. (2014) Analysis and experiment on a self-sensing ionic polymer–metal composite actuator. *Smart Materials and Structures*, **23**(7), 074007.

75. Jung, M. Y., Chuc, N. H., Kim, J. W., Koo, I. M., Jung, K. M., Lee, Y. K., Nam, J. D., Choi, H. R. and Koo, J. C. (2006) Fabrication and Characterization of Linear Motion Dielectric Elastomer Actuators. *Proceedings of SPIE 6168, Smart Structures and Materials 2006: Electroactive Polymer Actuators and Devices (EAPAD), 616824*, USA, doi: 10.1117/12.658145.

76. Chuc, N. H., Thuy, D. V., Park, J., Kim, D., Koo, J., Lee, Y., Nam, J.D and Choi, H. R. (2008) A dielectric elastomer actuator with self-sensing capability. *Proceedings of SPIE 6927, Electroactive Polymer Actuators and Devices (EAPAD) 2008, 69270V*, USA, doi: 10.1117/12.777900.

77. Rosset, S., O'Brien, B. M., Gisby, T., Xu, D., Shea, H. R., and Anderson, I. A. (2013) Self-sensing dielectric elastomer actuators in closed-loop operation. *Smart Materials and Structures*, **22**(10), 104018.

78. Chen, Z., Ren, W., Gao, L., Liu, B., Pei, S., and Cheng, H. M. (2011) Three-dimensional flexible and conductive interconnected graphene networks grown by chemical vapour deposition. *Nature Mat-*

erials, **10**(6), 424-428.

79. Xu, X., Li, H., Zhang, Q., Hu, H., Zhao, Z., Li, J., Qiao, Y and Gogotsi, Y. (2015) Self-sensing, ultralight, and conductive 3D graphene/iron oxide aerogel elastomer deformable in a magnetic field. *ACS Nano,* **9**(4), 3969-3977.

80. Jin, X., Deng, M., Kaps, S., Zhu, X., Hölken, I., Mess, K., Adelung, R. and Mishra, Y. K. (2014) Study of tetrapodal ZnO-PDMS composites: A comparison of fillers shapes in stiffness and hydrophobicity improvements. *PLoS One,* **9**(9), 0106991.

81. Raja, S. N., Zherebetskyy, D., Wu, S., Ercius, P., Powers, A., Olson, A. C., Du, D. X., Lin, L., Govindjee, S., Wang, L.W and Xu, T. (2016) Mechanisms of local stress sensing in multifunctional polymer films using fluorescent tetrapod nanocrystals. *Nano Letters,* **16**(8), 5060-5067.

82. Raja, S. N., Olson, A. C., Thorkelsson, K., Luong, A. J., Hsueh, L., Chang, G., Gludovatz, B., Lin, L., Xu, T., Ritchie, R. O. and Alivisatos, A. P. (2013) Tetrapod nanocrystals as fluorescent stress probes of electrospun nanocomposites. *Nano Letters,* **13**(8), 3915-3922.

83. Ciprari, D., Jacob, K., and Tannenbaum, R. (2006) Characterization of polymer nanocomposite interphase and its impact on mechanical properties. *Macromolecules,* **39**(19), 6565-6573.

84. Lehn, J. M. (2002) Toward complex matter: supramolecular chemistry and self-organization. *Proceedings of the National Academy of Sciences,* **99**(8), 4763-4768.

85. Rowan, S. J., Cantrill, S. J., Cousins, G. R., Sanders, J. K., and Stoddart, J. F. (2002) Dynamic covalent chemistry. *Angewandte Chemie International Edition,* **41**(6), 898-952.

86. Lehn, J. M. (2002) Supramolecular polymer chemistry-scope and perspectives. *Polymer International,* **51**(10), 825-839.

87. Giuseppone, N., and Lehn, J. M. (2004) Constitutional dynamic self-sensing in a zincII/polyiminofluorenes system. *Journal of the American Chemical Society,* **126**(37), 11448-11449.

Chapter 6

Advances in Flame Retardant Polymers

6.1 Introduction

6.1.1 Fire Determining Parameters

For the combustion process to start, three basic components must be present together, which are fuel, air/oxygen and energy/heat. This can be illustrated by a fire triangle demonstrated in Figure 6.1.

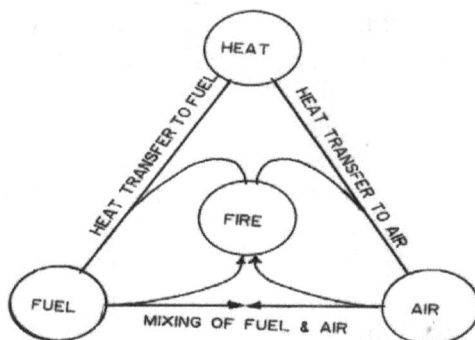

Figure 6.1 The fire triangle.

Fire is a highly complex process due to the multitude of variables affecting the basic components. It is, therefore, almost impossible to quantify and predict the actual course of fire. Energy can be transferred to the fuel by spark, flame and solar radiation. Continuation or progress of fire and its intensity depend on the availability of sufficient quantity of oxygen in the surrounding medium. The fuel itself influences the fire environment and the parameters of primary importance are shape and size, thickness, surface characteristics, specific heat and thermal conductivity as well as chemical properties, i.e., ignition, flash point and heat of combustion of the materials. The

Dipak K. Setua, Defence Materials & Stores Research & Development Establishment, DMSRDE (Post Office), G.T. Road, Kanpur 208013, India
E-mail: dksetua@rediffmail.com
© 2019 Central West Publishing, Australia

variety of parameters significant to fire performance suggests that it is not an intrinsic property of the material.

6.1.2 Course of Fire

The course of a fire can be split into several steps described below:

Initiation and Spread to Fully Developed Fire

As an ignition source ignites a combustible matter, heat is generated and further ignites the materials in the vicinity. The additional heat released and rise in temperature speed up the rate at which fire spreads in the surrounding. At one stage, the radiant heat and temperature are high enough the fire leads to decomposition of the material with evolution of flammable gases. The ignition of flammable gases causes an extremely high rate of fire spread in the entire area, and this point of time is called a "flash over". The fire can only be extinguished prior to the "flash over" stage. As the temperature exceeds 1000 °C and the entire vicinity of fire starts burning due to "flash over", it is termed as a fully developed fire. At this stage, the fire can no longer be controlled. With respect to polymers, the role of a flame retardant additive is to effectively reduce the influence of ignition point, fire spread and amount of heat release from the flammable polymers [1a]. Several phases of the course of a fire are also depicted in Figure 6.2 [1b].

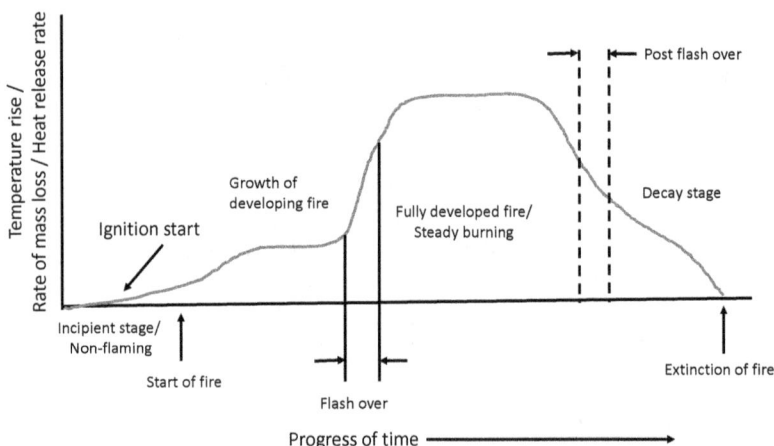

Figure 6.2 Phases in the course of a fire (adapted from Reference 1b).

The Burning of Polymer

The combustion of polymeric materials proceeds through various stages. A schematic diagram of the process is given in Figure 6.3 [1c].

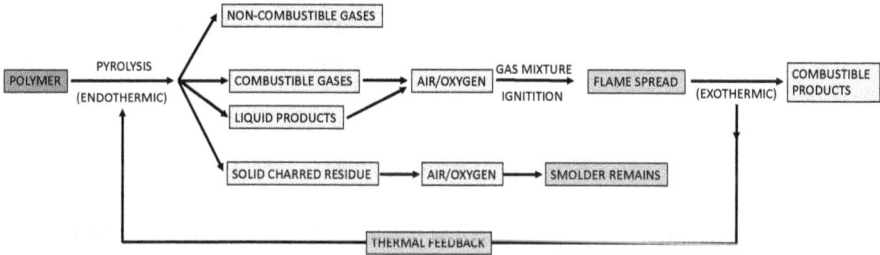

Figure 6.3 Various stages of the combustion process (schematic view) (adapted from Reference 1c).

Heating: A solid polymer is heated by thermal feedback or by an external heat source like radiation or flame. In the initial phase, the heat tends to soften or melt the macromolecular chains, e.g., thermoplastics. Thermosetting plastics and vulcanized rubbers have a three-dimensional crosslinked molecular structure, which prevents softening or melting. The polymers do not pass as such into the gas phase if further energy is not supplied, but decompose before vaporizing.

Decomposition: Decomposition or pyrolysis are endothermic processes in which sufficient energy must be provided to overcome the high binding energies of the bonds between individual atoms of polymer backbone and to provide necessary activation energy. Polymers differ in chemical properties, structure, decomposition temperature and thermal stability. Decomposition of polymers mostly precedes via free radical chain reactions initiated by traces of oxygen or other oxidizing impurities trapped during synthesis. In general, the resistance to thermal degradation depends on the chemical composition of the polymer. Polypropylene (PP), polyvinylchloride (PVC) and polybutadiene rubber (BR) are very susceptible to thermal degradation, whereas polymers such as polyether sulfone/polysulfone (PES/PSU), polyether-ether-ketone (PEEK) and polysiloxanes (silicones) possess excellent resistance to thermal and thermo-oxidative degradation due to the availability of strong sigma bonds in the long chain molecular backbone as well as the presence of different side-chain groups. The various stages of thermal-oxidative degradation of polymers are mentioned below.

Initiation: Oxidative degradation is usually initiated when polymer chains form radicals, either by hydrogen abstraction or by homolytic scission of a carbon-carbon bond. This can occur during synthesis, processing or service while exposing the polymer materials to light or heat.

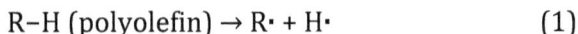

$$R-H \text{ (polyolefin)} \rightarrow R\cdot + H\cdot \qquad (1)$$

Propagation: The propagation of the degradation process involves a number of reactions. The first step is the formation of a free radical (R·), which forms a peroxy radical (ROO·) with an oxygen molecule (O_2). It subsequently abstracts a hydrogen atom from another polymer chain to form a hydroperoxide (ROOH). The hydroperoxide splits into two new free radicals (RO·) and (·OH), which abstract labile hydrogen from other polymer chains. Since two new free radicals are produced from each initiating radical, the process is accelerated depending on the ease of removal of hydrogen from other polymer chains and rate of free radicals termination via recombination and disproportionation mechanisms.

$$R\cdot + O_2 \rightarrow ROO\cdot \qquad (2)$$
$$ROO\cdot + RH \rightarrow R\cdot + ROOH \qquad (3)$$
$$ROOH \rightarrow RO\cdot + \cdot OH \qquad (4)$$
$$RO\cdot + RH \rightarrow R\cdot + ROH \qquad (5)$$
$$\cdot OH + RH \rightarrow R\cdot + H_2O \qquad (6)$$

Termination: Termination of degradation is achieved via recombination of two radicals or by disproportionation/hydrogen abstraction processes. Recombination of two chain radicals results in an increase of the molecular weight and crosslink density of the end-product, resulting in embrittlement and cracking of the polymer material.

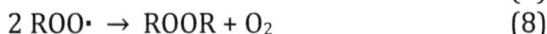

$$R\cdot + R\cdot \rightarrow R-R \qquad (7)$$
$$2 ROO\cdot \rightarrow ROOR + O_2 \qquad (8)$$
$$R\cdot + ROO\cdot \rightarrow ROOR \qquad (9)$$
$$R\cdot + RO\cdot \rightarrow ROR \qquad (10)$$
$$HO\cdot + ROO\cdot \rightarrow ROH + O_2 \qquad (11)$$

The termination by chain scission, on the other hand, results in a decrease in the molecular weight which leads to the softening of the polymer and reduction of the mechanical properties.

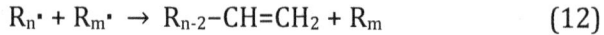

$$R_n{}^\bullet + R_m{}^\bullet \rightarrow R_{n-2}-CH=CH_2 + R_m \tag{12}$$
$$2\,RCOO^\bullet \rightarrow RC=O + ROH + O_2 \tag{13}$$

The relative predominance of the several termination steps depends on the structure of polymers and associated heat flux conditions. For example, PP and poly-isobutylene (PIB) with short alkyl side groups and unsaturated natural rubber (polyisoprene, NR) basically undergo chain scission, whereas polyethylene (PE), BR and polychloroprene rubber (CR) suffer from embrittlement due to crosslinking during thermo-oxidative breakdown. The above mentioned radicals can give rise to various decomposition products depending on the composition of the polymer, and essentially a variety of gaseous products is formed. In the case of polymethyl methacrylate (PMMA), the final result is depolymerization and formation of over 90% of the monomer species. On the other hand, the degradation of PE leads to the formation of both saturated and unsaturated hydrocarbons.

Ignition: The flammable gases formed during pyrolysis with atmospheric oxygen reach to the lower ignition limit and are either ignited by an external flame or if the temperature is sufficiently high, these can self-ignite [2]. The flash-ignition (FIT) and self-ignition (SIT) temperatures of various polymers/natural products are determined by ASTM D 1929 and are presented in Figure 6.4 [1c].

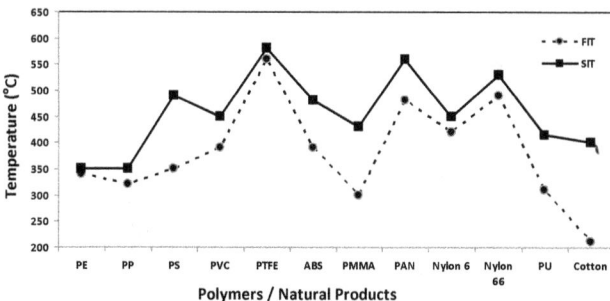

Figure 6.4 Typical FIT and SIT values of different polymers and natural products [1c].

Ignition depends on numerous variables such as abundance of oxygen, temperature as well as physical and chemical properties of polymers. The reaction of the combustible gases with oxygen is exothermic and, if sufficient energy is available, overrides some endothermic pyrolytic reactions and initiates further flame spread.

Flame spread: The exothermic combustion reaction reinforces the pyrolysis of the polymer by thermal feedback and fuels the flame at an increasing level. The progress of a hydrocarbon diffusion flame has following pathways [3]:

Growth:

$$CH_4 + OH \rightarrow CH_3 + H_2O \tag{14}$$
$$CH_4 + H \rightarrow CH_3 + H_2 \tag{15}$$
$$CH_3 + O \rightarrow CH_2 + OH \tag{16}$$
$$CH_2O + CH_3 \rightarrow CHO + CH_4 \tag{17}$$
$$CH_2O + H \rightarrow CHO + H_2 \tag{18}$$
$$CH_2O + OH \rightarrow CHO + H_2O \tag{19}$$
$$CH_2O + O \rightarrow CHO + OH \tag{20}$$
$$CHO \rightarrow CO + H \tag{21}$$
$$CO + OH \rightarrow CO_2 + H \tag{22}$$

Branching:

$$H + O_2 \rightarrow OH + O \tag{23}$$
$$O + H_2 \rightarrow OH + H \tag{24}$$

The chain branching is an extremely high energy step where H· and OH· radicals are formed and confer a high velocity to the progressive flame front [4]. A phenomenological description of the flame spread along a polymer surface is shown in Figure 6.5.

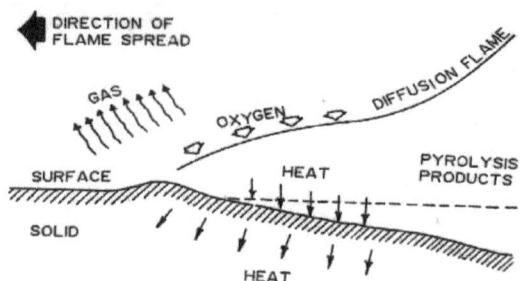

Figure 6.5 Flame spread (schematic view).

Another factor which determines the extent of flame spread is the heat of combustion of the polymer, as shown in Figure 6.6 [1c]. However, the heat of combustion should not be correlated to the combustibility of the individual material. Concurrent to diffusion controlled gas phase reactions which spread extremely rapidly, various other slower oxygen dependent reactions also take place concomitantly resulting in the production of smoke, soot and carbonlike residue (char) in the condensed phase in association with after flame luminescence, after glow, debris or incandescence.

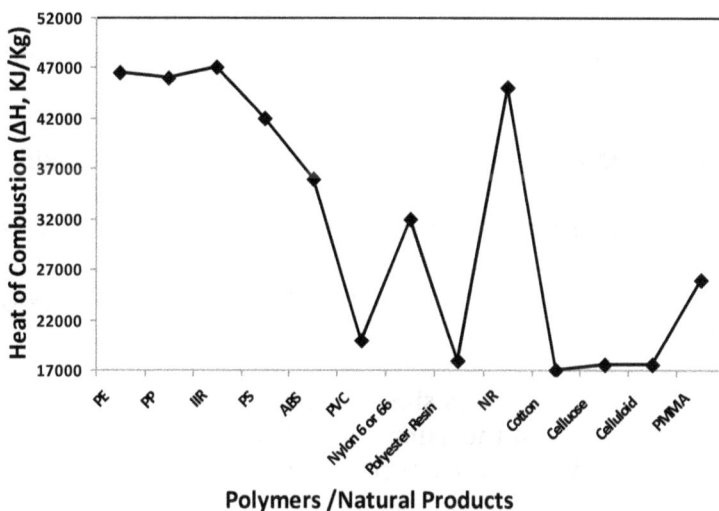

Figure 6.6 Heat of combustion of various polymers and natural products [1c].

6.2 Mechanism of Flame Retardation

On exposure to a sufficient heat flux for a long period, organic polymers undergo thermal degradation. In a large-scale fire or in a downward propagating candle like flame, very little oxygen reaches to the polymer surface, where the nature and amount of the evolved gases are controlled by the pyrolytic thermal degradation. Consequently, the susceptibility of the polymer to thermal degradation is determined by its chemical composition and also affected by the presence of filler, pigments, stabilizers and plasticizers, along with the type and concentration of flame retardants [5]. Different polymers generate different types of volatile products in varying concentrations over distinct temperature ranges. The chemical steps leading to the generation of volatiles may differ as per the electronic structure of the decomposed group [6-7]. Three basic schemes, which are common to most of the polymer degradation processes, are described as follows:

1. Random chain cleavage followed by chain unzipping, which are characterized by high monomer yield and a slow decrease in molecular weight of the polymer.

2. Random chain cleavage in steps, characterized by low monomer yield and a rapid drop in the molecular weight.

3. An intra-chain chemical reaction followed by cross-linking and formation of carbonaceous residue along with chain cleavage. This process generates a relatively high yield of volatiles from inter-chain reaction, very small extent of monomer as well as negligible loss of molecular weight of the chains during the initial stages of polymer degradation.

In some cases, there occurs simultaneously more than one reaction scheme depending on the sample size, heating rate, pyrolysis temperature, environment and nature of additives.

6.3 Flame Retardant Strategies

A flame retardant interferes with the physics and chemistry of polymer combustion in a preferred environment in such a way that the flammability properties are improved significantly [8]. An effective flame retardant additive must have the following criteria:

1. The additive or its decomposition products are intended to volatilize simultaneously with the evolved gases generated by thermal degradation of the substrate and inhibit the vapor phase combustion of the fuel gases.

2. The flame retardant is intended to alter the path of thermal degradation by providing a low energy process that promotes solid state reaction leading to the carbonization over materials generating combustible gases.

3. The flame retardant also occasionally forms a protective coating that insulates the substrate from the thermal energy that promotes degradation and evolution of fuel gases.

6.3.1 Solid State Chemistry

Continued generation of fuel gases through thermal degradation of polymers is a critical criterion for a sustained flaming. Therefore, the most effective strategy for flammability reduction would be to interfere with the mechanism of degradation in such a way so as to prevent formation of volatile fuel gases. Flame retardants which act in this manner are referred to as condensed phase active material. Examples are phosphorus containing compounds like phosphoric acid, phosphorous salts, organo-phosphorous compounds (e.g., phosphates and phosphonates), etc. Effective use of condensed phase flame retardants requires that the polymeric material leads to the formation of a non-volatile residue at the pyrolysis temperature, as in the case of cellulose [9].

6.3.2 Thermal Barrier

Generation of volatile gases can also be prevented by protecting the surface from the flame by creation of a thermal barrier called intumescent coating. As this coating comes in contact with sufficient thermal energy, it softens by melting or chemical decomposition and strongly acidic catalysts are activated. These catalysts accelerate the decomposition of other coating components, resulting in the generation of non-combustible gases, which act as a blowing agent for the softened coating. During the final stages of intumescences, the viscosity of the foamed mass increases owing to cross-linking reactions and the coating is further stabilized [10].

6.3.3 Thermal Sinks

In conventional polymer formulations, fillers are used to strengthen the mechanical properties as reinforcement, improve the processing characteristics or to reduce the compound cost. These fillers can also function as thermal sinks. In filled compounds, the fillers, mostly inorganic, are added in amounts about 30-40 phr of polymer. However, only the addition of inert fillers is not sufficient to meet the flammability requirements since very high filler loading would be required which can't be afforded for technical reasons. Examples of such fillers are talc, calcium carbonate, crushed marble clay, titanium dioxide, etc. Hydrated alumina is specifically used for this purpose. Energy of the flame, which pyrolyzes the polymer and generates fuel gases, is consumed by the endothermic dehydration of alumina by removal of water of vaporization from the surface of the filler. The water vapor also absorbs energy from the flame as it passes through its front.

As in case of intumescing coating, the thermal sinks do not alter the inherent combustibility or solid state chemistry of the base polymers.

6.3.4 Vapor-phase Chemistry

In case either coating or thermal sink are not sufficient to improve the flammability of polymers by solid-state chemistry, flames are inhibited by active retardants which create interference with the vapor-phase chemistry of polymers subjected to combustion. Basically, these flame retardants reduce the flow of thermal energy from the flame back to the solid surface. The simplest way to lower heat flux from the flame to the solid is to introduce non-reactive species into the flame as in case of water elimination from the hydrated alumina. However, the non-reactive species generated by the endothermic pyrolysis of these additives also act as inert diluents in the flame functioning like non-combustible gases.

Another mechanism of perturbation of the flame chemistry is to reduce the heat generation by the combustion reaction. Carbon monoxide is promoted over carbon dioxide formation to interact with the H·, O:, HO· radicals, which are responsible for the continued propagation of the flame [11-12]. The organo-halogen compounds employed as flame retardant additives in thermoplastics act in this way. In addition, metal ions, metal oxides and phosphorous compounds are also capable to inhibit flame propagation through similar mechanism [13].

Vapor- phase-active flame retardants are equally effective in various polymers over wide range of temperatures and follow a completely different mechanism with halogen inhibition reactions. However, in the context of the efficiency to decrease the proximity of a combustion process, the vapor-phase active ingredients are comparatively less effective than the condensed phase active species [14].

6.3.5 Smoldering

Porous materials may undergo a different mode of combustion. Unlike flaming, smoldering is a surface phenomenon and mostly occurs in open cell foam cushions, e.g. in case of flexible polyurethane foams [15]. It is a slow process which often takes hours to run its course, whereas a flaming might consume the same mass in minutes. Similar to smoldering is glowing combustion, where the material shows a visible thermally induced incandescence. Afterglow refers to glowing combustion that follows the extinction of flaming combustion. It does not require that the initial source be porous, but rather the residue may be of char like. Smoldering can be initiated spontaneously by self-heating or bringing in an ignition source. In either case, endothermic decomposition of the porous substrate leaves a porous and active char of high radical density. Oxygen slowly diffuses through the char, often against the outgoing flow of pyrolysis gases, and reacts with the active sites of the flammable material producing carbon monoxide and releasing energy. The effect of smoldering to a person residing in a close vicinity of the material is the inhalation of a substantial amount of CO from a combustion producing relatively less heat. Effective flame inhibitors are not necessarily effective smolders and glow inhibitors and vice versa. For example, sodium borate and sodium molybdate are effective flame retardants for cellulose, but these have little effect on smoldering. Similarly, boric acid and ammonium borates inhibit smoldering, but these are ineffective toward flame retardation. Sodium borate and sodium molybdate are effective char formers, but the sodium cation accelerates smoldering. As a general class, phosphorous compounds are effective. Occasionally, sulphur is also used to inhibit smoldering.

6.4 Flame Retardant Additives

6.4.1 Antimony Compounds

Antimony trioxide (Sb_2O_3) has been used as a white pigment since ancient times. It behaves as a condensed phase flame retardant in cellulosic materials. It can be applied by impregnating a fabric with a soluble antimony salt followed by second treatment that precipitates in the fibers. As the treated fabric is exposed to flame, the oxide reacts with the hydroxyl groups of the cellulose (fabric) causing them to decompose endothermally. The decomposition products, water and char cool the flame reaction while slowing down the extent of flammable decomposition products.

Sb_2O_3 and antimony pentoxide (Sb_2O_5) are the most widely used commercial flame retardants. This is because of their synergistic interaction with organo-halogens. However, Sb_2O_3 alone has no effect on the flammability of butyl rubber compositions [16]. Sb_2O_3/halogen based flame retardant systems tend to increase the smoke production during combustion, which is a health and environmental issue for possible toxicity and carcinogenicity [17,18]. Melamine complexes of bismuth or antimony trichloride and tribromide are thermally stable up to 200-230 °C as well as chemically stable to withstand processing conditions for most polymers. Their fire retardation ability is much greater than a synergistic mixture of the corresponding metal oxide/organo-halogen combination. Thermal behavior of these complexes shows a delayed volatilization effect of metal halide which causes an extended protection of flammability of polymers by condensed and gas phase mechanism [19].

6.4.2 Alumina Trihydrate

Alumina trihydrate, $Al(OH)_3.10H_2O$ (ATH), is also used as a secondary flame retardant and smoke suppressant for flexible PVC and polyolefin formulations, in combination with antimony and other halogen additives. Addition of a minor amount of either zinc borate or phosphorous compounds results in the formation of glasses which insulate the unburned polymer from flame [20]. ATH is the least expensive but also less effective as a flame retarder as its effectiveness is one-fourth as compared to conventional halogen based flame retardants. Usually a high loading about 50-60% of ATH with respect to polymer is needed in order to obtain an effective flame retardancy and its application is limited to polymers whose processing temperatures do not exceed 220 °C.

On exposing to temperatures above 280 °C, ATH loses its water of hydration. The reaction is strongly endothermic, consumes thermal and radiant energy from the flame and slows the rate of pyrolysis of the substrate. In addition, the vaporized water acts as a diluent and cools the flame, thus, reducing the effective heat flux to the substrate surface. The alumina residue obtained in the process acts as a thermal shield on the substrate surface. Moreover, it does not promote incomplete combustion, reduces toxic gases and acts as a smoke suppressant. ATH is used in applications where its high loading in polymers is permitted, e.g. upholstery coatings, thermoset polyesters and reinforced unsaturated polyesters [21].

6.4.3 Antimony-Halogen Synergism

Antimony oxide shows excellent synergistic effect with chlorinated and brominated organic flame retardants and provides an effective flame retardant system [22]. Antimony and halogens react at flame temperature to form corresponding trihalide or oxyhalide depending on the mole ratio of the reactants and structure of the organic compounds. Tribromide or trichloride are formed directly when the mole ratio of halogen to antimony is at least 3:1 and the halogen compound undergoes dehydrogenation as

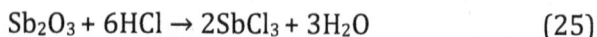

$$Sb_2O_3 + 6HCl \rightarrow 2SbCl_3 + 3H_2O \qquad (25)$$

If the mole ratio of halogens is less than 3, there can't be dehydrohalogenation and oxyhalide is formed as

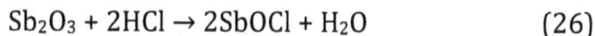

$$Sb_2O_3 + 2HCl \rightarrow 2SbOCl + H_2O \qquad (26)$$

The formation of $SbCl_3$ from SbOCl has been described by two mechanisms. One asserts that the formation by thermal disproportionation as

$$5SbOCl \rightarrow Sb_4O_5Cl_2 + SbCl_3 \qquad (27)$$

The other possibility is the formation of $SbCl_3$ via further chlorination of the oxyhalide. In case the organo-halogen compound does not form acid halide easily, a third compound capable of reacting with antimony and halogen is needed. This can be the polymer itself wherein the polymer decomposes before the organo-halogen to generate a free radical, which reacts with the labile hydrogen of the polymer to form hydrogen halide and a new polymeric radical. This hydrogen halide can react with Sb_2O_3 and generate $SbCl_3$. In the vapor phase, $SbCl_3$ undergoes free radical decomposition in step-wise manner, enabling the chlorine to remain in the flame zone longer than it would have if it was generated directly from the organic halogen compounds and, therefore, is more effective. The fine antimony mist formed by decomposition of $SbCl_3$ also participates in flame inhibition by deactivating oxygen, hydrogen and hydroxyl radicals.

6.4.4 Boron Compounds

Boron compounds are well known for imparting flame retardancy to polymeric materials. The most widely used is zinc borate. It is prepared as an insoluble double salt from water soluble zinc and compounds with varying amount of zinc, boron and water of hydration are available. The ratio of these affects the temperature of flame-inhibition as well as temperature at which the polymers can be processed [23]. Zinc borates can either be used alone or in combination with other halogen synergists, e.g. Sb_2O_3. Occasionally, ATH is also used to form a glass like substance that inhibits polymer degradation. Barium metaborate is used as flame retardant and antifungal agent for many flexible PVC compounds. Boric acid and sodium borate, known as borax, are used in cellulosic composition for flame retardancy, however, these do not represent a durable solution. Ammonium fluoroborate, NH_4BF_4, is unique as it generates both halogen and boron based flame retardant. Sb_2O_3 is recommended as a synergist. Boron is considered both as condensed and vapor-phase flame retardant. Under flaming conditions, boron and halogens form corresponding boron trihalides which are effective Lewis acids. They promote crosslinking which minimizes decomposition of polymers into volatile flammable gases. These trihalides are also volatile, thus, vaporize into the flame, release halogen and function as flame inhibitors.

6.4.5 Magnesium Hydroxide

Magnesium hydroxide, $Mg(OH)_2$, is another metal hydrate that decomposes endothermally like ATH, accompanied by the formation of H_2O. Its decomposition temperature is 300 °C, which is 100 °C higher than ATH. Therefore, it is used for polymers requiring higher processing temperature. Magnesium hydroxide has been reported as a smoke suppressant and an effective flame retardant in polypropylene as well as elastomers [24,25].

6.4.6 Molybdenum Compounds

Molybdenum oxides are being used to impart flame resistance and smoke suppression to plastics [26,27]. Molybdic oxide, ammonium octamolybdate and zinc molybdate are the most widely used molybdenum based flame retardants. Molybdenum trioxide is a condensed phase flame retardant, and it decomposes to produce non-volatile products which tend to increase the char yield. Molybdenum compounds are recommended for PVC, its alloys and unsaturated polyesters. Low loading of molybdic oxide increases the char yield significantly in plasticized PVC from 9.9 to 23.5%, and molybdenum is subsequently recovered from char when burned completely. Molybdenum trioxide also acts as smoke suppressant by promoting formation of *cis* rather than *trans* polymeric decomposition products which are precursors for smoke generation.

6.4.7 Tin Compounds

Inorganic tin compounds including tin oxide (SnO_2) have been used as flame retardants for cellulose since 18th century. Since 1970s, these are used as a synergist for halogen compounds in the same way as antimony oxides. Anhydrous zinc stannate, zinc hydroxy stannate and SnO_2 are the most important tin based flame retardants and can act as potential replacements for Sb_2O_3 in halogen containing polymers including PVC, polyester and alkyl resins [28]. Zinc stannate is stable up to 530 °C, as compared to zinc hydroxyl stannate and stannic acid which decompose approximately at 180 °C. In association with chlorinated paraffin wax, SnO_2 and zinc hydroxyl stannate, at low level, confer increased flame retardancy and reduced rate of smoke evolution. Flame retardation mechanism of tin, although not fully understood, follows both condensed phase as well as vapor phase inhibition reaction, where chlorinated paraffin wax causes extensive volatilization of tin. Flame retardant behavior is observed in the absence of halogen synergist, which is a marked difference from antimony compounds.

6.4.8 Organic Flame Retardant Additives

Chlorine and bromine containing organo-halides are commercially significant as flame retardants for rubbers and polymers [29]. Fluorine compounds are occasionally ineffective, and iodine compounds are too expensive. Generally brominated aliphatics, chlorinated aliphatics and brominated aromatics are used, and their thermal stability follows the order: brominated aromatics > chlorinated aliphatic > brominated aliphatics. In aliphatic halogen compounds, thermal stabilizers are added. Brominated aromatics are used in thermoplastics at fairly high temperature without a heat stabilizer. The flame retardant should decompose to liberate halogen at somewhat lower temperature than decomposition of the polymer. Most of the brominated additives do not react chemically with the polymer, but some are of reactive type, e.g. tetrabromo bisphenol A (TBBPA). Among varieties of brominated diphenyl oxides, referred to as diphenyl ethers, a few important materials are described in the following:

Decabromo Diphyenyl Oxide (DBBO)

It is a thermally stable material, suitable for polymers which require high processing temperature [30]. It is always used in combination with Sb_2O_3. Many commodity and engineering polymers, elastomers, thermosetting resins, textiles and latex based coatings use this flame retardant additive. However, with styrene resins, UV stabilizers are required to avoid discoloration.

Other Bromo-compounds

Octabromo diphenyl oxide is used with acrylonitrile butadiene styrene (ABS) thermoplastic polymer which has poor UV stability. Pentabromodiphenyl oxide is used with flexible polyurethane foams. On blending with triaryl phosphate, it generates superior thermal stability and scorch resistance. Bis-tribromophenoxy ethane (BTBPE) possesses better thermal stability and is primarily used with ABS polymer. TBBPA is a reactive flame retardant which is used in ABS to obtain high flow and better impact properties. It is more UV stable than decabromo diphenyl oxide but inferior to BTBPE in performance.

Ethylene bistetrabromo phthalimide is a specialty product used in engineering thermoplastics and polyolefins. It is thermally stable, has resistance to bloom and provides UV stability to styrene resins. Hexabromo cyclodecane, obtained from bromination of cyclic oligomers of butadiene, possess up to 74.7% bromine and is mainly used with styrene foams. Bromination of styrene generates poly(dibromo styrene). Similarly, brominated epoxy oligomers and brominated carbonate oligomers are other oligomeric flame retardants commonly used with polyamides and polyolefins. Examples of some other reactive flame retardants are tetrabromophthalic anhydride, dibromo neopentyl glycol, vinyl bromide and brominated phenols.

Chlorinated Flame Retardants

Chlorinated paraffin with varied molecular weight of the starting paraffin from 12-24 carbons and chlorine content varying from 40-70 wt% in the final product are marketed as solid flame retardant additives. Besides, liquid chlorinated paraffin is also used both as a plasticizer as well as a flame retardant. Bishexachloro cyclopentatieno cycloctane possesses fairly good thermal stability like brominated compounds.

Phosphorus-containing Flame Retardants

Halogen-containing flame retardants act in the gas phase, whereas the phosphorus based retardants mainly influence the gaseous decomposition of the polymers in the condensed phase. The flame retardant is converted into phosphoric acid by thermal decomposition and extracts water from the pyrolyzing substrate causing it to char. Lyons [31] as well as Rabe *et al.* [32] have described the condensed reaction steps as following:

1. The non-volatile polymer-phosphoric acid inhibits progress of pyrolysis of polymers by forming a carbonaceous char layer as a glassy coating on the surface of polymer particles preventing further degradation by oxygen and radiant heat.

2. Extensive charring is promoted by the formation of compounds like phosphines, CO and CO_2 in the gas phase, which are reducing agents, and results in suppression of afterglow of the solid substrate.

There is also the evidence of phosphorous-containing flame retardants which are effective in the gas phase. Hastie and McBee [33] have reported on formation of compounds triphenyl-phosphine oxide [$(C_6H_5)_3PO$] (TPPO) which break down into fragments, as per eq. 28-30, in the gas phase. The HPO catalyzes the recombination of H atoms to give rise to H_2 (equation 31), thus reducing the flame energy.

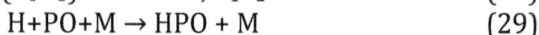

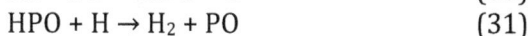

$$(C_6H_5)3PO \rightarrow PO, P_1P_2 \tag{28}$$
$$H+PO+M \rightarrow HPO + M \tag{29}$$
$$OH + PO \rightarrow HPO + O \tag{30}$$
$$HPO + H \rightarrow H_2 + PO \tag{31}$$

Phosphorus - Halogen Synergism

Various assumptions pertaining to the synergistic effect of phospho-rus-halogen combinations as flame retardant are as following:

1. Phosphorus halides and oxyhalides are good coolers and better radical interceptors than hydrogen halides (HX). In addition, they have high boiling points and higher specific gravities than HX and, therefore, remain longer in the reaction zone. This increases the probability of enhanced reaction with the radicals of fuel gases sup-porting the flame.

2. Halogen-containing flame retardants do not give off their entire halogen as HX in the gas phase, whereas the phosphorus-halogen combination generates HX quantitatively because the P-X bond is weaker than C-X bond.

3. On account of their higher molecular weight, the phosphorous halides form a layer over the condensed polymeric phase, thus, shielding it from oxygen required for combustion.

These assumptions have been supported by Weil [34] and Levchik and Well [35]. However, whether the presence of phospho-rous-halogen combined phase exerts any major influence on the combustion of fuel gases or reduced proximity of combustion through a condensed phase char formation, as evidenced in case of only phosphorus types of compounds, are still unclear.

Phosphorus-Nitrogen Synergism

The synergistic effect of phosphorous and nitrogen compounds on engineering plastics have been reported [36]. In cellulose, these fa-cilitate the phosphoration of cellulose with phosphoric acid. This accelerates the dehydration of cellulose and leads to charring of the substrate [37]. Further, it has also been observed that there occurs 1) a charred zone which is covered by liquid polyphosphoric acid, 2) the formation of a glassy layer of heat resistant polymers, and 3) the formation of cross-linked polyphosphazenes. However, except in the case of cellulose, it is believed that the synergistic effect predomi-nates in the condensed phase, similar to phosphorous compounds used alone.

The nitrogen compounds can promote the formation of a layer of porous charcoal by gaseous decomposition which acts as a heat shield of the combustion zone. Besides, the nitrogen compounds prevent the phosphorous compounds from escaping by pyrolysis into the gas phase, where they are less effective than in the condensed phase. The nitrogen compounds themselves form acids like HNO_2, HNO, etc., which lead to the dehydration of the polymers via a carbonium ion mechanism, and these compounds on pyrolysis form volatiles which reach the gas phase to act as free radical interceptors.

6.5 Flammability of Elastomers

Ignition and burning characteristics of elastomers become a matter of concern in a few specific applications and, therefore, the sequence of processes of heat degradation and combustion are interesting [38-40]. Non-combustible gases generally serve to dilute the flammable gases and, in many cases, generate a non-burning gas blanket, thus, preventing oxygen to reach the combustion zone. The burning of an elastomer is essentially controlled by the amount of the pyrolytic gases as well as the rate at which these gases are formed under given conditions. Any carbonaceous char formed in the burning process, particularly if it possesses a good coherence and structural integrity, also renders a stable insulating layer which protects underlying elastomer from a sustained combustion.

The thermal degradation of the elastomers occurs at relatively mild temperatures and is accompanied by the evolution of a large quantities of volatile and highly flammable hydrocarbon materials, which burn to generate significant amount of heat. The calorific value of hydrocarbon compounds and all-organic polymers is high as compared to inorganic and hybrid inorganic-organic polymers. The evolved heat on combustion is available to cause further pyrolysis. Therefore, the rubber once ignited burns vigorously and needs to be protected to decrease their potential contribution to fire hazards.

In developing a flame-retardant formulation, care should be taken not to impair any of the important physical properties of the rubber. Generally, an optimum formulation is worked out empirically, and, in many cases, compromises are made between performance, flame retardation and cost. It is also necessary to establish, for any specific case, a frame of reference with respect to the degree of flame retardancy desired as well as the type of flammability test, simulating the actual service condition. Three approaches are basically used to arrive at an optimum rubber composition, e.g. type and concentration of additives, co-polymerization of starting rubber and post curing treatment. For example, SBR and other rubbers are rendered flame retardant by the addition of a variety of organic halogen compounds, e.g. chloroparaffin (CP), chlorinated polyethylene elastomer (CPE), butadiene-vinylidene chloride copolymers, CR or organo-bromo phosphates preferably used in combination with antimony trioxide, hydrated alumina, zinc and calcium borates. NBRs are compounded with a variety of halogenated additives, e.g. PVC and chloroparaffin [41]. BR can be protected by a combination of organobromides and a radical generating compound such as N-nitroso-N- methylaniline or N- nitroso-cabazole [42].

Influence of different flame retardant additives, e.g. combination of DBBO and Sb_2O_3, ATH, CP, and zinc hydroxystannate (ZHS), on thermal, mechanical, flammability behavior, besides cost effectiveness of NR, have been reported [43]. 10 phr of ZHS in combination with 90 phr ATH and 40 phr of CP was found to have the best balance of fire retardance for NR. Increase in the flame retardant capability of NR (like synthetic CR) by the addition of halogenated free ZHS and halogenated flame retardants (DBBO, CP), in combination with ATH and Sb_2O_3, has further been supported by Patarapaiboolchai and Chaiyaphate [44]. In regard to the corrosiveness and toxicity of smoke during combustion of NR, ZHS has been reported to be an ideal alternative to Sb_2O_3 [45]. ZHS loaded at 20 phr in combination with 40 phr CP showed better flammability performance and generated lower smoke density than Sb_2O_3. However, partial replacement of flame retardant additives in NR by ZHS produced no synergistic improvement. The pyrolysis of ethylene-co-propylene (EPM) rubber and ethylene-propylene-diene monomer (EPDM) rubber has been studied by Perejon *et al.* [46]. The authors found that the activation energy for the pyrolysis of EPM decreased with increasing propylene content, and the thermal stability of the EPDM rubber depended on the composition of diene monomer. A detailed treatment of the mechanism of pyrolysis of PIB and other polyolefins has been reported by Madrosky and Strauss [47]. The authors analyzed the evolved volatiles from pyrolysis of these polymers by mass spectroscopy and observed the presence of monomers of the respective polymers. Differential thermal analysis (DTA) revealed the occurrence of both exo- and endothermic degradation. Further, it was reported that the presence of quaternary carbon atoms in PIB in butyl rubber (IIR) also reduced the thermal stability [48]. Smith and King [49] studied ignition parameters of NR and found that the surface temperature of radiation heated test specimen by a pilot flame falls between 316-502 °C, and auto ignition took place at surface temperature 620-670 °C. Straus and Madorasky [50] studied clay filled and peroxide cured BR samples with different extent of curing and hardness, e.g. 45, 60 and 94 Shore A. On subjecting to thermogravimetric analysis (TGA), the samples were observed to start losing weight at 420-430 °C, with completion of degradation at 485-495 °C and final weight loss for all the samples between 93-98%. Similarly, NR also exhibited similar thermal stability for both raw rubber and vulcanizates with different level of crosslinking.

In case of thermoplastic polyurethane (TPU) elastomer, which is widely used in TPU coated textiles and high performance fabrics, Kutty and Nando [51] have investigated the effect of various flame retardant additives, such as halogenated polymers, Sb_2O_3/chlorine donors' combination, zinc borate and ATH [51]. Smoke generation was found to reduce significantly, while limiting oxygen index (LOI) reduced marginally by addition of short Kevlar fibers. The maximum improvement of LOI was in case of Sb_2O_3/CP combination in weight ratio 1:6, and 70 phr loading of ATH improved LOI associated with substantial reduction of smoke generation. The authors also observed that TPU dripped down during burning in a flame, and degradation of TPU took place into polyols and isocyanates and further into low molecular weight species. This was supported by a two-stage thermal degradation peaks of TPU, one at 383 °C and other at 448 °C by thermo-gravimetric analysis. In case of TPU coated Kevlar, an increase in the thermal stability of the polymer coupled with a smoke density reduction by nearly 50% were reported.

6.5.1 Flame-retardant Rubber Coated Textiles: A Case Study

Rubber is used in combination with textiles, commonly to make these waterproof and airtight, along with achieving environmental protection. In fact, rubber and textile technologies have developed more or less in parallel, leading to products which meet stringent property and performance requirements for many applications. These include inflatable boats, air cushions and airbeds, fuel tanks and bladders, balloons, parachutes, inflatable boats, protective clothing, etc. The fabrics can be used for strategic applications in defense for a variety of military activities (readers interested in more details are referred to the reference text 1 mentioned at the end of the chapter). Illustrations of some of these are provided in Figure 6.7. These are used for protective clothing and equipment to counter nuclear, biological and chemical (NBC) threats, as well as to provide flame resistance, weatherproofing and cold-resistant garments for high altitude and snowbound areas.

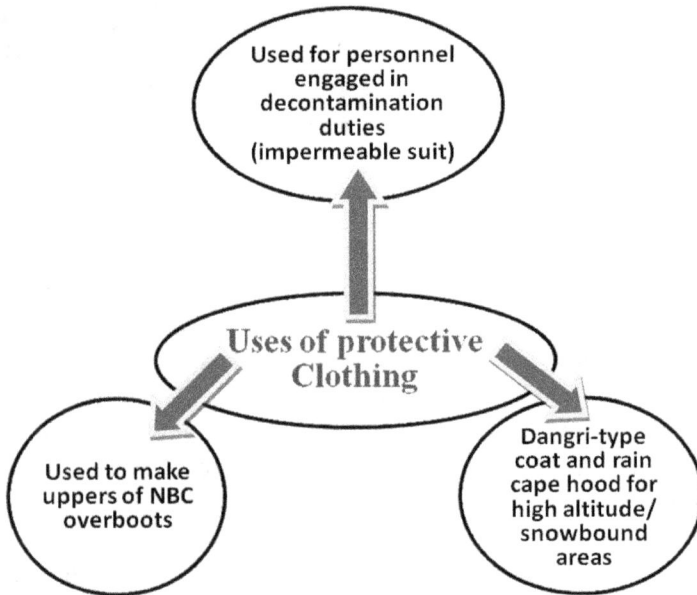

Figure 6.7 Military uses of protective clothing.

Application of rubber-containing coatings requires specialized coating and processing techniques.

Preparation and Characterization of Coated Fabrics

For this purpose, the industry commonly employs two methods described below:

Dry method: It involves the calendering process, where a rubber is compounded in a mixing mill or Banbury mixer. A three or four roll rubber calendar is used to apply the dry rubber to textile substrate. The rolls of the calendar form the rubber into the required thickness and subsequently combine it to the textile fabric. The rubber compound can be applied to the fabric in any desired weight.

Solution or cement method: Rubber compound is dissolved in a solvent (toluene or MEK) and spread upon a textile substrate. Only a small amount of coating is applied by the solution method, and the heavier coatings are applied by the dry method.

For this, the compound material is cut into small pieces and dissolved in a churn, to 35% solids. Application of rubber to the textile is achieved by impregnation or a spreading machine, which is simply a knife over a rubber roller with ovens attached. The fabric is pulled through the gap between the knife and rubber roll, and through the oven. The temperatures of the ovens vary, depending upon the type of the solvents and rubber, into zones of 150, 200 and 275 °F. This type of oven is known as three-zone oven. Inflatable safety equipment, such as life-vests, life-boats and escape suits, which are carried aboard aircraft, provide the proofing industry with a group of products requiring low coating weight and high strength.

Curing: Curing or vulcanization of rubber coated fabric is necessary to provide effective physical properties to the rubber compound and is accomplished by festooning the fabric in a dry heat chamber under specific conditions of time and temperature as per specific requirements of the finished goods. Coated fabric is also vulcanized in a roll form, where it is dusted with mica or talc and wrapped around a metal drum. Water resistant cover is put over the drum, and the total assembly is rolled into a closed chamber of a steam vulcanizer.

Materials and testing methods: Wide choice of substrates, e.g. cotton, nylon, polyester, fiberglass and blends of these fabrics are used. The employed elastomers include natural rubber, styrene type rubber, nitrile, neoprene, hypalon, polysulfide, butyl, silicone and polyacrylic types. Woven fabrics are treated before use by a boiling-off process which removes all the finishes. These are then treated on a contact hot drum at 160 to 190 °C in order to minimize the shrinkage at processing temperatures. A few basic tests (based on ASTM procedures) commonly performed are tensile, tear, adhesion of coating to fabric, solvent, oil and hydrostatic resistance, etc. Strategic applications of these coated fabric or garments require further specialized coatings and processing techniques to modify the base fabric in order to impart flame retardancy, oil and water repellency and sufficient air permeability [16]. Apart from rubber, other materials are also occasionally applied, e.g. PVC, PU, water based PU, etc.

Determination of Flammability Performance

Flammability can be tested for vertical flammability and LOI as per BS 3119 and ASTM-D-2863/77. In the vertical flammability test, a test specimen of size (50.8 x 300 mm) and thickness up to 1.2 mm is exposed to open flame for 10s, the flame is removed and flammability characteristics of the material are determined as follows:

After flame: The length of time for which the material continues to flame after the ignition sources is removed.

After glow: The time for which the material continue to glow, i.e. absence of flaming but with emission of light from combustion zone.

Char length: The maximum extent of damage or charring of the material measured in the vertical direction, ignoring any surface effects such as scorching or smoke deposition.

The details of LOI test method are also presented in the subsequent section. Burning characteristics of textile fibers are summarized in Table 6.1.

High Resolution Thermogravimetry (HR-TGA) Coupled Mass Spectrometry

ASTM procedures (D 568-61, D 635-63, D 757-49, D 1692-571 and D 1360-58) can be used to evaluate the flammability of materials. However, these methods have disadvantages of being empirical and time-consuming. Burning of rubber generates fuel for the flame and provides energy to sustain the pyrolytic reactions and flame propagation. Therefore, an accurate estimation of the evolved gaseous materials is important. Evolved gas analysis (EGA) can be performed by coupling TGA with a mass spectrometer (MS). For kinetic studies, traditionally both isothermal and non-isothermal methods are applied. However, experiments with a constant heating rate are faster than a series of isothermal experiments. In conventional TGA, the requirement of high resolution necessitates a low heating rate and, therefore, an increased experimental time. For a balance between the experimental time and high resolution, HR-TGA has been developed [52]. In this method, the heating rate is dynamically varied to maximize the resolution and to separate closely occurring events.

Table 6.1 Burning characteristics of the textile fibers

Fiber type	*Burn or melt*	Shrink in flame	Residue	Other properties
Acetate	Burns & melts	Yes	Dark, hard, solid, bead	Acid odor (hot vinegar type)
Acrylic	Burns & melts	Yes	Hard, irregularly shaped bead	Flame gives off black smoke; acid odor
Aramid	Burns & melts	Yes	Hard, black bead	Self-extinguishing
Cotton	Burns	No	Fine, feathery grey ash	Odor similar to burning of paper
Flax, hemp, jute, ramie	Burn	No	Fine, feathery grey ash	Odor similar to burning of paper
Glass	Melts at high temperature	Very slowly	Hard mass	Flame resistant; heat from lighter does not cause melting of the fiber
Nylon	Burns & melts	Yes	Hard, cream colored bead; if fibers are overheated, bead will become dark	Flaming, usually by the finishes present; drops of molten fiber fall from heated portion
Olefin	Burns & melts	Yes	Hard, tan bead	Flame gives off black smoke; chemical odor
Polyester	Burns & melts	Yes	Hard, cream colored bead; if fibers are overheated, bead will become dark	Drops of molten fiber fall from heated portion of the sample; flame gives off black smoke; chemical odor
Rayon	Burns rapidly and melts	No	Fine, feathery grey ash	

Rubber Coated Textiles for Snow Clothing, Extreme Cold Climate and NBC Applications

The effect of various chemical additives (e.g. brominated phenyl ether (decabromo diphenyl oxide)), chlorinated paraffin wax, inorganic oxides (e.g., Sb_2O_3) and halogenated rubber (e.g. polychloroprene rubber (CR)) on the flame retardancy behavior of brominated isobutylene-co-isoprene rubber (BIIR) has been studied. The decomposition profile, evolved gases and kinetics of decomposition process at a dynamic heating rate with HR-TGA coupled with MS were used. The HR-TGA results were compared with results from conventional thermogravimetric analysis (TGA) at a constant heating rate. A lifetime and temperature relationship was evaluated using Toop's method, and it was observed that the shelf life decreased sharply with temperature.

Nylon fabric coated with CR and BIIR blends has been studied. Several elastomeric blends of CR and BIIR were prepared in blend ratio of 40:60 (CR:BIIR) and converted into dough in toluene, which were coated on both sides of nylon fabric. The flame retardancy of the fabric was enahnced by adding 30 parts of decabromo diphenyl oxide. Five parts of chlorinated polyethylene (CPE) rubber was added as compatibilizer. These fabrics can be used for the fabrication of clothing equipment for NBC applications in defense as well as garments for high altitude/snow bound areas with integrated hood. Besides, Setua *et al.* [53] have also evaluated the bio-deterioration characteristics of these fabrics by thermal analysis and evaluated other functional properties affected by bio-exposure simulating to actual service conditions for use of service personnel.

6.6 Flame Retardancy of Other Polymeric Materials

In a fire incident, smoke and toxic gases (STG) are more hazardous than the fire itself. Principal hazards of STG are obscuration and poisoning. Obscuration is caused by particulate materials suspended in the air. Poisoning depends on the quality and quantity of the toxic materials in the combustion products. Thus, reducing smoke generation in the fire is as important as improving flame resistance of the materials [54].

In general, flame retardancy in polymeric materials is achieved by the incorporation of inorganic oxides of alkaline earth metals, halogenated organic materials and elastomeric additives. Fire occurs when an ignition breaks down polymer strands, thus, creating chemical entities which are vaporized. At sufficiently high temperature, these fragments react with oxygen in air to release more heat, thus, resulting in breaking of more polymeric strands yielding fuel gases. The fire becomes fatal when the heat dispersed by the material exceeds the marginal enthalpy required to produce a steady stream of vapor phase. Less fire prone products generally have either inherently more stable polymeric structure or contain fire retardant additives.

Fire retardant additives are often used to improve fire performance of low to moderate cost commodity polymers. These additives may be physically blended or chemically bonded with the host polymer. These generally affect either lower ignition susceptibility or lower flammability once ignited. Ignition resistance can be improved solely from thermal behavior of the additives in the condensed phase. Retardants such as hydrated alumina add to the heat capacity of the product, and the endothermic volatilization of bound water can be a significant component of effectiveness of this family of retardants. Other additives, such as organophosphates, change the decomposition chemistry by the formation of cross-linked and more stable solid as well as lead to the formation of a surface char layer. This layer insulates the product from further thermal degradation and impedes the flow of potentially flammable decomposition products from interior of the product to the gas phase where combustion would occur.

Flame retardancy can also be imparted to polymers by incorporating elements such as bromine, chlorine, antimony, tin, molybdenum, phosphorous, aluminum, magnesium, etc., during the generation of polymer compounds. Phosphorus, bromine and chlorine are usually incorporated as organic compounds. Addition of at least 40 wt% of the halogen based flame retardants is generally required to impart a reasonable degree of flame retardancy. However, at the same time, these adversely affect the mechanical properties of the polymer and eventually generate acidic gases (e.g. hazardous HCl). Halogen containing species vaporize at the same temperature as the polymer fragments and co-exist in the reactive area of the flame. Halogens are effective at decreasing the concentration of free radicals propagating the flame, thus, reducing the flame intensity and burning rate. For a small ignition source, the use of flame retardants as such can produce self-extinguishment. More intense flame source may overwhelm the flame retardant necessitating a higher concentration of the additive. ATH is used alone because it is not synergistic with the halogens and is useful in applications where halogens are barred or in cases where a low processing temperature for polymers is used. Various aspects of thermal stability, synergistic effect of combination of flame retardants in fire retardant polymers including thermoset matrix composites and textiles, etc., have been discussed by several authors [55-60].

6.7 Flame Retardant Polymer Nanocomposites

Polymer nanocomposites exhibit remarkably improved flame-retardancy and are environmentally friendly as these do not produce toxic volatiles on flaming. Reduced flammability of polymer nanocomposites is achieved by the incorporation of nanofillers, especially silicates (clays), inorganic hydroxides, carbonaceous materials (e.g. carbon nanotubes and graphene), metal oxides and polysilsequioxanes. The characteristic size of these fillers is typically in the range of 1-100 nm. Occasionally, the combination of different nanofiller systems offers synergism, and their surface modification by chemical methods or surfactant improves their distribution and dispersion. This enables an exfoliated structure in the polymer matrix system, which promotes extraordinary flame retardancy. The controlling factors of flame retardant mechanisms for polymer nanocomposites are the physical barrier effect rendered by the nanofillers (with different morphologies, e.g. 0D, 1D, 2D structures), extent of char formation (as barrier to O_2, fumes and thermal isolation), radical trapping, peak heat-release rate (HRR) and formation of 3D nanostructure as in case of inorganic-organic or in metal-organic framework systems.

Concurrent to the enhancement in fire resistance by addition of a small concentration of nanofillers (typically 1-5 wt%) as compared to pure polymers, mechanical properties, gas permeability, thermal stability, softening temperature, thermal conductivity and functional properties (electrical, optical and magnetic) can be varied by adding various particulate fillers. For instance, the optical and magnetic properties of nanocomposites can be varied over a wide range and optimized for a particular application, e.g. optical clarity with UV blocking effect are important for packaging and camouflage materials used in defense, along with the development of active armor, climate control materials for extreme heat, cold, rain and wind as well as communication devices (GPS integrated optical/wireless types). The nanocomposites also exhibit antifungal activity for the development of sophisticated health-care medical devices and drug delivery systems for bio-medical sectors.

Another important factor for the industrial sector is overall cost-performance benefit of the nanomaterials. It is important to note that the nanocomposites achieve significant improvement in the desired properties at a low concentration of filler. Also, the polymer nanocomposites can be manufactured using existing process technology, e.g. a conventional epoxy resin containing nanoparticles can be transfer molded by injection of resin into a fiber lay-up. If additional functional properties can be imparted by the choice of nanomaterials, it may also be possible to eliminate the need of additional materials added for combination of properties, e.g. in automotive sector.

The key drivers for the use of polymer/elastomer nanocomposites in the automotive industry are reduction in vehicle weight, improvement in engine efficiency (fuel saving), enhanced fire safety and reduction in CO_2 emissions, along with other safety features and driving comfort (a detailed discussion about the usage of a variety of nanofillers (synthetic and natural) can be found in reference text 2 mentioned at the end of the chapter). Thermal stability and flame retardancy for polymer blends, composites and nanocomposites based on rubber, thermosets and thermoplastics essentially containing surface modified layered silicates, alumina, fullerenes, LDH, etc., as well as the role, chemistry and mechanism of flame retardancy and potential applications are also detailed in reference texts 3 and 4.

Scope of development of a heat resistant and non-flammable general purpose rubber (based on polyisoprene SKI-3 + polybutadiene SKD rubber compound with a sulphur vulcanization system) by partial replacement of antimony trioxide with organoclays (Cloisite 10A, Cloisite 15A and Cloisite 30B montmorillonites) and combination of aluminum hydroxide and barium borate has been reported on by Petrova *et al.* [61]. Significant reduction of flammability in chloro-isobutylene isoprene rubber (CIIR) has also been achieved by the addition of a low concentration (approx. 3 phr) of multilayer graphene (MLG) using conventional rubber processing in two-roll milling [62]. Addition of at 180 phr of ATH was observed to generate 66% reduction in peak heat release rate, 160% increase in ignition period and 69% reduction of smoke density in thermoplastic elastomer based on EPDM and low-density polyethylene (LDPE), as studied using a cone calorimeter [63]. Dendrimer modified organo-montmorillonite (DOMt) has also been used as flame retardant additive in NR compounds and the effect of its concentration on the mechanical performance, thermal stability and flame-retardant properties has been studies [64]. At 20 wt% loading of DOMt, the flammability parameters of NR, such as heat release rate (HRR), smoke evolution area (SEA) and carbon monoxide (CO) concentration were observed to be significantly reduced, as compared to unfilled NR [64]. Flammability reduction along with decreased heat release rate for polypropylene-graft maleic anhydride and polystyrene-layered-silicate nanocomposites using montmorillonite and fluorohectorite have been studied by cone calorimetry [65]. The type of silicate, nano-dispersion and degradation process were reported to influence the extent of flammability properties. This observation is further supported by the thermal stability of PS nanocomposites generated using iron-containing clay and graphite [66]. Though intercalation of iron containing clay in PS exhibited enhanced thermal stability as compared to exfoliated systems, PS-graphite nanocomposites both with and without iron exhibited no significant difference as iron was not nano-dispersed in the graphite. Thermal and flammability properties of highly filled polyamide 6 (PA6)/organo-clay nanocomposites have also been reported on by Dasari *et al.* [67]. The nanocomposites exhibited effective filler dispersion and flame retardation as compared to PP or polybutylene terephthalate (PBT) nanocomposites prepared by conventional polymer melt extrusion technique. The observation has also been confirmed by X-ray diffraction (XRD) and transmission electron microscopy (TEM) [68]. Various perspectives of polymer nanocomposites in respect to their combustion, flame retardant properties including HRR and smoke emission (SEA), performance evaluation tests (such as oxygen index and UL-94 classification), synergistic effects of combination of nano-fillers (layered silicates, carbon nanotube, LDH, etc.) as well as future development of flame retarded polymer nanocomposites are widely available in the published literature [69-73]. Some recent studies in this area proposed the development of polymeric flame retardant nanocontainers (NOFLAME – 705054) with high thermal stability, low flammability, good dispersion in polymeric matrices as well as ability for encapsulation of a wide range of substances, which makes these nanocontainers attractive to many futuristic applications [74]. In another study, a synergistic improvement in heat release reduction with formation of a strengthened char by a blend of multiwall carbon nanotube (MWCNT) mixed with organoclays for development of a flame retardant and insulated wire and cable compound based on PE has been reported [75]. A comprehensive review has also been published on the electrical, mechanical and thermal properties of crosslinked polyethylene (XLPE) and epoxy resin based micro/nanocomposites containing different commercial nanofiller and varying in filler size, type and distribution in polymer matrix. The use of these micro- and nanocomposites as insulation systems in high-voltage engineering, such as cables, generators, motors, transformers etc., has been found to be promising as compared to traditionally used insulator materials [76]. The influence of layered compounds, mainly montmorillonite (MMT), layered double hydroxide (LDH), layered metal phosphate and layered carbon (e.g. graphene) on the flammability performance of polymer nanocomposites has also been reported [77]. The barrier effect of these layered compounds is responsible for superior flame retardancy, however, these cannot decrease the total heat release (THR) of the nanocomposites alone. However, these layered compounds in combination with a traditional flame retardant (either metal ions or phosphorous based flame retardant additives) can catalyze the charring process during combustion and restrain the release of combustible gases and are, thus, found to be more effective for fire safety of the polymeric materials.

6.8 Flammability Testing

A limitation associated with flammability rating is the lack of clear and uniform definition of flammability of materials. American Society for Testing and Materials (ASTM) lists over one hundred methods for assessment of flammability ranging from small scale ignitability of the material to the case of an actual full scale fire test. In this respect, an overview of polymer flammability from a materials science perspective and test methods to quantify the fire behavior, along with mathematical relationship between polymer properties and chemical structure has been reported [78]. Specific test methods and regulations to interpret flammability of textile fabrics relevant to their many fold applications have also been described by Nazare and Horrocks [79]. Fire safety aspects of polymeric materials used commercially, nature of by-products in combustion and environmental safety norms, etc., have also been cited in a series of reports compiled by National Materials Advisory Board, National Research

Council, USA [80]. The commonly used test methods are listed in Table 6.2 and described as follows:

6.8.1 Material Tests

These tests are separate from the actual measurement of flammability on the products.

LOI

The minimum concentration of oxygen in an O_2/N_2 mixture that supports combustion of a vertically mounted test specimen of size 0.65 x 0.3 x 12.5 cm^3 is called limiting oxygen index. The combustion of the sample should continue for at least 3 min or the flame should cover a distance of 50 mm. The main advantage of this test is its reproducibility, which makes it useful for quality control. However, the disadvantage is that the results rarely correlate with the results of other flammability tests.

Specific Tests

Federal Motor Vehicle Safety Standard (MVSS), USA used this test to measure the burning behavior of materials used in automobile interiors. A specimen of size 35.5 x 10.1 x 1.3 cm^3 is mounted horizontally and ignited for 15 s. The burning rate should be below 10 cm/min, however, the auto-makers typically impose more severe criteria.

The Underwriters Laboratory UL-94 Standard for Safety measures the ignitability of plastics. The test specimens of size 12.7 x 1.27cm^2 and thickness varied from 3.2, 1.6 and 0.8 mm are mounted vertically and ignited using a Bunsen burner held at an angle of 30° for 10 s. If the specimen self-extinguishes after the first ignition, further 10 s cycle is applied for second ignition. A layer of cotton is placed under the test specimen to collect the flaming drips. The flammability is classified as (1) V-0, if no specimen burns longer than 10 s and a total of five specimens are tested. Two flame applications per specimen and 10 number of total tests should not last more than 50 s. Nevertheless, the underneath cotton should also not ignite in any case; (2) V-1, applies for no specimen burning longer than 30 s. The sum of the after-flame times for five specimens (two flame applications per specimen and total 10 number of flame applications) must

Table 6.2 Flammability tests methods

Designation	Description	Characteristics measured
ASTM E162-87	Surface flammability of materials using a radiant heat energy source.	Flame spread
ASTM E119-88	Test methods for fire tests of building construction materials	Fire endurance
MVSS 302	Flammability testing materials for automotive parts e.g., polyurethane foam	Burning rate
ASTM D2863-87	Oxygen index	Measuring the minimum oxygen concentration to support candle like combustion: extinction
ASTM E662-83	Smoke production in fires: small-scale experiments	Smoke type, concentration
ASTM E84 (UL 723)	Surface burning characteristics for materials	Comparison of flame spread and smoke developed values for product materials
UL 94, UL 790, ASTM E 108-90	Broad range of fire tests that yield a wide variety of ratings and test values	The UL 94 (HB, V-0, V-1, V-2) ratings are used to classify the flammability of polymeric materials used for parts in devices and appliances/ignition resistance
	Roof burn	Flame spread, Class A, B, or C fire rating for a roof assembly
UL 1715	Standard for fire test of interior finish material	Classification of interior finish material assemblies by use of a standardized room fire exposure.

CAL 133, CAL 117	California Technical Bulletin 133 (abbreviated CAL 133) is a regulation for the upholstered furniture industry, just like CAL 117 method	Spread of fire through highly flammable materials like foams, fabrics and other materials
ASTM E1353-16, ASTM E1354	Standard test methods for cigarette ignition resistance of upholstered furniture/heat and visible smoke release rates using an oxygen consumption calorimeter equipment	Ignitability, heat release rate, heat of combustion, etc., for materials and products
ASTM E906	Standard test method for heat and visible smoke release rates using a Thermopile method, cone calorimeter, OSU heat release rate calorimeter	Heat release and smoke density for materials and products
Factory Mutual Flammability Test	A large-scale test for building products like wall and ceiling. FMRC flammability apparatus sometime used as alternate to this test.	Heat, smoke, toxic and corrosive products release, ignition and flame spread

not exceed 250 s. Also, the cotton cannot be ignited; and (3) V-2, the test method is same as V-1 except the cotton can be ignited. Most commercial plastics meet either V-0 or V-2 classifications.

Smoke Generation

Smoke generation during burning is estimated as per ASTM procedure D 2843-77 in a smoke density test chamber. A light source is placed at the top of the chamber such that light traverses across the whole chamber space. A detector, also placed at the top, is separated from the source by glass windows. The detector output is adjusted to 0% absorption or 100% transmittance when the sample is not burning. The detector output is also compensated for variation of temperature in the environment of the cabinet.

The test samples are exposed to a standard flame on a wire gauge at the bottom of the chamber. The gas pressure as well as the flame length are controlled as the sample continues to burn. The generated smoke fills-up the chamber and obscures the passage of light. The readings are taken at interval of 30 s after the sample is exposed to flame, and the maximum value obtained is reported as the smoke density.

Heat Release Calorimeters

There are three principal types of heat release calorimeters. The cone calorimeter measures the rate of heat release of a burning test specimen of size 10 x 10 cm² and up to 2.5 cm thickness exposed to a radiant heat up to 100 kW/m². Many parameters in addition to heat release rate may be measured, e.g. total heat release, mass loss, time to ignition, critical heat flux and smoke production. The amount of heat released is calculated based on the consumption of oxygen and, therefore, it is also sometimes termed as oxygen consumption calorimetry. The advantage of this test is that materials can be subjected to heat fluxes similar to those encountered in real fires, but the correlation is needed between the laboratory test results and actual large scale firing scenario.

The Ohio State University, USA has developed another type of calorimeter as a true adiabatic instrument which measures heat released during burning of polymers by measurement of the temperature of the exhaust gases. This test method has been adopted by the Federal Aeronautics Administration (FAA), USA to test total and peak heat release of materials used in the interiors of commercial aircraft.

The other principal heat release test developed by the Factory Mutual Research Corporation (FMRC) is a flammability apparatus, different from above mentioned calorimeters, which allows measurement of flame spread as well as heat release and smoke. A unique feature of the FMRC system is that it uses oxygen at concentration higher than ambient to simulate reversed radiations from the flames of a large-scale fire.

6.8.2 Product Tests

Tests in which the finished articles are subjected to a more or less realistic fire are called product tests. A few examples are as follows:

Tunnel Test

The tunnel test is a widely used method for testing the flame spread potential of building products such as electrical cable and wall coverings. The test apparatus consists of a tunnel of 7.62 x 0.445 x 0.305 m³ in cross-section, one end of which holds two gas burners. The total heat supplied by the burners is 5.3 MJ/min. The test specimen of size 7.62 x 50.8 cm², attached to the ceiling, is exposed to the gas flame for 10 min, while the maximum flame spread, temperature, and evolved smoke are measured.

Factory Mutual Corner Test

This is a large-scale corner test used to test building products. The test rig consists of three sides of a cube. The two walls are 15.24 and 11.58 m by 7.62 m tall. The ceiling is 9. 14 x 15.24 m². The product to be tested is mounted on the walls and ceilings in a manner consistent with the intended use. The fire source is a 340 kg stack of wood pallets located in the corner. In order to pass the test, no flame can propagate to any extremity of the walls or ceiling. The Factory Mutual flammability apparatus is proposed to replace this test for certain applications.

CAL 133

California Technical Bulletin 133 is a test of the fire hazard associated with upholstered furniture. The test is carried out by igniting a standard fire source directly on the piece of furniture being tested. In the recent version of the test, the fire source is a gas flame. Smoke, heat and toxic gas emissions are measured during the test. A related test, BS 5852, uses various wooden cribs as the fire source.

Flame Spread

Flame spread of a polymer film can be measured by means of CSI flammability tester. This method is also applicable to textile fabrics and paper. Flame spread characteristics of a material can be evaluated in terms of the time required for ignition, rate of burning and the temperature of flame produced on burning. The test specimen dimension is 76 x 228 mm². The method follows ASTM D1439-77 standard. The burning rate is calculated in mm/sec and average time of burning is also calculated.

An important time saving and confidence building measure in the vast jargon of testing methods adopted for the evaluation of polymer flammability or flame retardancy is the simulation and modeling approach. Zhang *et al.* [81] reported a mathematical model for the burning process of fire-retardant intumescent PP. The proposed model was observed to appreciably predict the mass loss rates and temperature distribution in the material during the course of fire. Efficacy of an inverse modeling approach in case of pyrolysis of polymers (e.g. PMMA) has been claimed to resolve the complexity of a large number of experimental parameters involved in solid ignition, flame spread and fire growth, along with the difficulty to quantify their values from direct measurements [82]. In the inverse modeling approach, five models are assumed with a number of parameters ranging from 3 to 30, and the equations as well as experimental data are coupled. Similar levels of accuracy with strong compensation effects between implemented mechanism and model complexities were observed. Pyrolysis and flammability of PMMA by transient irradiation has also been modeled, the 1D model coded in GPyro, by Vermesi *et al.* [83]. The predictions were compared to experimental results by cone calorimeter for constant irradiation as well as fire propagation apparatus for transient irradiation.

6.9 Summary and Conclusions

An ideal flame retardant should be easy to incorporate in the polymers, should not bloom or bleed out on storage and should not significantly alter the mechanical properties. Furthermore, it should be colorless, exhibit good stability to UV and visible light, be resistant to ageing and hydrolysis, etc. It should also have matching decomposition temperature with the polymers and remain effective throughout the decomposition period. It must be non-corrosive, odorless, non-toxic, effective in small concentration, smoke resistant and cheap.

A number of formulations are required to be developed for the optimization of mechanical and fire properties, especially in the presence of commonly used fillers and other additives. In case the flame retardant liberates acids on decomposition, e.g. hydrogen halide from halogen based compounds and phosphoric acid from phosphorous containing compounds, these can cause corrosion of the processing equipment, which are counteracted by adding suitable stabilizers. Flame retardants used on a commercial scale in plastics include organic chlorine and bromine compounds, phosphorous compounds or phosphorous compounds in association with chlorine and bromine compounds. Synergists such as antimony trioxide, organic radical initiators and inorganic compounds such as aluminum hydroxide, which act both as a flame retardant as well as a filler, are commonly used. Plastics containing flame retardants with aromatically bonded bromine, which decompose at high temperature, are easier to process, however, it is less effective than the aliphatic-bonded bromine compounds. In order to improve the efficacy of aromatic type flame retardants, synergists like antimony trioxide are used. Phosphorous based flame retardants also influence the physical properties of the polymers by acting as plasticizer.

Additive types of flame retardants are exclusively used in pure carbon chain structured polymers like polyolefins, polyvinyl chloride and polystyrene as well as for polymers with a heterogeneous chain structure such as polyurethanes, polyesters and polyamides. These are cheap and easily processable with the polymers. However, reactive types of flame retardants are preferred for their scope of evolution of specific formulation with particular set of properties and accomplishment of a durable polymeric product without any migrated or bleed-out non-volatile residue.

The principal chlorine-containing organic flame retardants are chlorinated hydrocarbons and chlorinated cycloaliphatics. These materials have low cost and good light stability, however, there are disadvantages due to the requirement of a large quantity for desirable level of flame retardancy, which adversely affects the mechanical properties of plastics. Commonly used aliphatic compounds are chloroparaffins, which are available in liquid as well as in solid form depending on the chlorine content between 30 to 70%. These have semi-plasticizing properties and poor thermal resistance. The most widely used cycloaliphatic chlorine compounds are hexachloroendomethylene tetrahydrophthalic acid (HET) and its anhydride, which are stable up to 260 °C. Aromatic chlorine compounds, although thermally stable up to 280 °C, are less prominent because of their poor flame retardant action.

Organic bromine compounds are more popular than chlorine compounds since these are required in lower concentration, have non-bleeding nature and are readily processable in plastics, however, these exhibit poor stability to light and are expensve. These materials are used as both reactive as well as additive types of flame retardants. Amongst the additive aliphatic bromine compounds, hexabromobutene is the most effective, though it has poor thermal stability. Cycloaliphatics, such as hexabromocyclododecane and pentabromochlorocyclohexane, are preferred for commercial use, as these offer superior flame retardation and better temperature resistance than linear aliphatic compounds. Aromatic bromine compounds, such as decabromodiphenyl ether, have breakdown temperature above 400 °C and are used in rubbers and plastics. Their flame retardant potential can be further improved by using these along with antimony trioxide. The most frequently used reactive type flame retardant containing bromine are TBBA, tetrabromophthalic anhydride and dibromoneopentyl glycol.

Compared to organic chlorine or bromine compounds, phosphorous based flame retardants are versatile as phosphorous exists in several oxidation states. Thus, phosphines, phosphine oxides, phosphonates and phosphates, including elemental red phosphorous, are used as flame retardants. Frequently, phosphorous compounds also contain halogens, in particular bromine, which increases the effectiveness of the flame retardant. Many of the phosphorous compounds are liquids and possess plasticizing properties, e.g. phosphoric acid esters, aryl phosphates and their alkyl-substituted derivatives. These are commercially employed in thermoplastics, in particular in PVC. Tris(dibromopropyl) phosphate is a highly effective flame retardant, however, its use is not permitted due to its mutagenic and carcinogenic nature. A large range of reactive flame retardants of phosphorous compounds is also available. Usually polyols modified with phosphorous based phosphines and phosphonates are commonly used. Other reactive types of phosphorous modified compounds include isocyanates, acids, nitrogen-containing and unsaturated compounds, etc.

Reference Texts

Setua, D. K. (2012) Emerging technology & fabrics: Polymer coated textiles with smart, intelligent and nano-structured fibers. In: *Polymer Coated Textiles and Recent Advances*, Akovali, G. (ed.), iSmithers/Rapra, UK, pp. 315-359.

Setua, D. K., and Gupta, Y. N. (2018) Elastomer-clay nanocomposites with reference to their automobile applications and shape-memory properties. In: *Rubber Nanocomposites and Nanotextiles in Automobiles*, Banerjee, B. (ed.), iSmithers/Rapra, UK, pp. 135-173.

Flame Retardants: Polymer Blends, Composites and Nanocomposites (Engin-

eering Materials), Visakh, P. M., and Arao, Y. (eds.), Springer, USA (2015).

Thermally Stable and Flame Retardant Polymer Nanocomposites, Mittal, V. (ed.), Cambridge University Press, UK (2011).

References

1. a) Hull, T. R., and Stec, A. A. (2009) Polymers and fire. In: *Fire Retardancy of Polymers: New Strategies and Mechanisms*, Hull, T. R., and Kandola, B. K. (eds.), Royal Society of Chemistry, UK, pp. 1-14.
   b) *Flame Retardant Additives for Coatings, Plastics, and Additives*, presentation slide of Walter Conti, Buckman Lab., USA.
   c) *Flame Retardant Polyolefins*, Smarttech Global Solution Ltd., India (2018).
2. *Performance Properties of Plastics and Elastomers, Handbook of Polymer Science and Technology*, Cheremisinoff, N. P. (ed.), 2nd volume, Marcel Dekker, USA (1989).
3. *The Mechanisms of Pyrolysis, Oxidation, and Burning of Organic Materials*, Wall, L. A. (ed.), National Bureau of Standards, USA (1972).
4. Aseeva, R. M., and Zaikov, G. E. (1985) Flammability of polymeric materials. In: *Key Polymers Properties and Performance, Advances in Polymer Science*, Springer, Germany, pp. 171-229.
5. Madorasky, S. L. (1975) *Thermal Degradation of Organic Polymers*, R. E. Krieger Pub. Co., USA.
6. *Polymer Additives in Stabilization: Performance and Mechanisms*, Grassie, N., and Chien, J. C. W. (eds.), Polymer Degradation and Stability, volume 20, pp. 179-366 (1988).
7. Morgan, A. B., and Gilman, J. W. (2013) An overview of flame Retardancy of polymeric materials: Application, technology, and future directions. *Fire and Materials*, **37**, 259-279.
8. *Organic Coating and Plastics Chemistry*, Craver, J. K., and Tess, R. W. (eds.), American Chemical Society, USA (1975).
9. *Cellulose Chemistry and its Applications*, Nevell, T. P., and Zeronian, S. H. (eds.), Halsted Press, John Wiley, USA (1985).
10. *Flame Retardant Polymeric Materials*, Lewin, M., Atlas, S. M., and Pearce, E. M. (eds.), Springer, USA (1975).
11. Hastie, J. W. (1973) Molecular-basis of flame inhibition. *Journal of Research of the National Bureau of Standards - A. Physics and Chemistry*, **77A**, 733-754.
12. Creitz, E. C. (1970) Literature survey of the chemistry of flame inhibition. *Journal of Research of the National Bureau of Standards - A. Physics and Chemistry*, **74A**, 521-530.
13. Miller, D. R., Evers, R. L., and Skiner, G. B. (1963) Effects of various inhibitors on hydrogen-air flame speeds. *Combustion and Flame*, **7**,

137-142.

14. Yeh, K., Drews, M. J., and Barker, R. H. (1980) Calorimetric study of polyester/cotton blend fabrics. V. Characterization of flame retardant action, *Journal of Fire Retardant Chemistry*, **7**(2).

15. Ohlemiller, T. J., Bellan, J., and Rogers, F. (1979) A model of smoldering combustion applied to flexible polyurethane foams. *Combustion and Flame*, **36**, 197-215.

16. Setua, D. K., Pandey, A. K., Debnath, K. K., and Mathur, G. N. (1999) Flame retardant and impermeable rubber coated fabric. *Kaut. Gum. Ku.*, **52**(7-8), 486-492.

17. Lewin, M. (1999) Synergistic and catalytic effects in flame retardancy of polymeric materials - An overview. *Journal of Fire Sciences*, **17**(1), 3-19.

18. Growth, D .H., Steller, L. E., Burg, J. R., Busey, W. M., Grant, G. C., and Wong, L. (1986) Carcinogenic effects of antimony trioxide and antimony ore concentrate in rats. *Journal of Toxicology and Environmental Health*, **18**(4), 607-626.

19. Touval, I. (2004) Flame retardants, antimony and other inorganic agents. In: *Kirk-Othmer Encyclopedia of Chemical Technology*, Scidel, A. (ed.), volume 11, John Wiley & Sons Inc., USA, pp. 1-19.

20. Ning, Y., and Guo, S. (2000) Flame-retardant and smoke-supressant properties of zinc borate and aluminium trihydrate-filled rigid PVC. *Journal of Applied Polymer Science*, **77**(14), 3119-3127.

21. Paul, S. (2010) Flame Retardant Additive for Polymers, Free of Halogens, Antimony Oxide and Phosphorous Containing Substances, patent US2010/0004370A1.

22. Babushok, V. I., Deglmann, P., Kramer, R., and Linteris, G. T. (2017) Influence of antimony-halogen additives on flame propagation. *Combustion Science and Technology*, **189**(2), 290-311.

23. Eltepe, H. E. (2004) The Development of Zinc Borate Production Process, MS Thesis, Izmir Institute of Technology, Turkey.

24. Larcey, P. A., Redfern, J. P., and Bell, G. M. (1995) Studies on magnesium hydroxide in polypropylene using simultaneous TG-DSC. *Fire and Materials*, **19**(6), 283-285.

25. Holloway, L. R. (1988) Application of magnesium hydroxide as a flame retardant and smoke suppressant in elastomers. *Rubber Chemistry and Technology*, **61**(2), 186-193.

26. Liu, L., Wu, W., Xue, H., and Qu, H. (2013) A series of metal molybdates as flame-retardants and smoke suppressants for flexible PVC. *Advanced Materials Research*, **634-638**, 1881-1885.

27. Moore, F. W., Weber, T. R., and Tsigdinos, G. A. (1981) Advances in the use of molybdenum compounds as smoke suppressant for PVC. *Journal of Vinyl & Additive Technology*, **3**(2), 139-142.

28. *Plastics Additives: An A-Z Reference*, Pritchard, G. (ed.), Springer-

Science, Netherlands (1998).

29. Boryniec, S., and Przygocki, W. (2001) Polymer combustion pro-
    cesses. 3. Flame retardants for polymeric materials. *Progress in
    Rubber and Plastic Technology*, **17**(2), 127-148.

30. *Decabromodiphenyl   Oxide*,   ScienceDirect,   USA.   Online:
    https://www.sciencedirect.com/topics/chemistry/decabromodip
    henyl-oxide [accessed 19th September 2018] (2018).

31. Lyons, J. W. (1970) *The Chemistry and Uses of Fire Retardants*,
    Wiley-Interscience, USA.

32. Rabe, S., Chuenban, Y., and Schartel, B. (2017) Exploring the modes
    of action of phosphorus-based flame retardants in polymeric sys-
    tems. *Materials*, **10**(5), 455.

33. Hastie, J. W., and McBee, C. L. (1975) *Mechanistic Studies of Tri-
    phenylphosphine Oxide – Poly(ethyleneterephthalate) and Related
    Flame Retardant Systems*, U. S. Government Publishing Office, USA.
    Online: https://www.gpo.gov/fdsys/pkg/GOVPUB-C13-
    50513e7c684c09c7d229f04823626afd/content-detail.html   [ac-
    cessed 20th September 2018].

34. Weil, E. D. (1978) Phosphorus-based flame retardants. In: *Flame -
    Retardant Polymeric Materials*, Lewin, M., Atlas S. M., and Pearce E.
    M. (eds.), Springer, USA, pp. 103-131.

35. Levchik, S. V., and Well, E. D. (2006) A review of recent progress in
    phosphorus-based flame retardants. *Journal of Fire Sciences*, **24**(5),
    345-364.

36. Leu, T. S., and Wang, C. S. (2004) Synergistic effect of a phosphorus
    - nitrogen flame retardant on engineering plastics. *Journal of Ap-
    plied Polymer Science*, **92**(1), 410-417.

37. Sharma, N. K., Verma, C. S., Chariar, V. M., and Prasad, R. (2015)
    Eco-friendly flame-retardant treatments for cellulosic green build-
    ing materials. *Indoor and Built Environment*, **24**(3), 422-432.

38. Fabris, H. J., and Sommer, J. G. (1977) Flammability of elastomeric
    materials. *Rubber Chemistry and Technology*, **50**(3), 523-569.

39. Janowska, G., Jastrzabek, A. K., Rybinski, P., Wesolek, D., and
    Wojcik, I. (2010) Flammability of diene rubbers. *Journal of Ther-
    mal Analysis and Calorimetry*, **102**(3), 1043-1049.

40. Nishizawa, H. (2006) Flame retardant agents. *Nippon Gomu Kyo-
    kaishi*, **79**(6), 316-322.

41. Whelan, W. P. (1977) Flame Retarded NBR/PVC Compositions,
    patent US4043958.

42. Lyon, R. E. (2013) *Plastics and Rubber*, McGraw Hill, USA. Online:
    https://polymerandfire.files.wordpress.com/2013/03/chapter-3-
    plastics-and-rubber1.pdf [accessed 18th September 2018].

43. Hassim, D., Harper, J. F., and Ansarifar, A. (2008) Influence of flame
    retardant additives on the flammability behaviour of natural rub-
    ber (NR). *Journal of Rubber Research*, **11**(4), 223-236.

44. Patarapaiboolchai, O., and Chaiyaphate, S. (2010) Improvements of natural rubber for flame resistance. *Songklanakarin Journal of Science and Technology*, **32**(3), 299-305.

45. Kind,. D. J. (2012) A review of candidate fire retardants for polyisoprene. *Polymer Degradation and Stability*, **97**(3), 201-213.

46. Perejon, A., Sanchez-Jimenez, P. E., Gil-Gonzalez, E., Perez-Maqueda, l. A., and Criado, M. (2013) Pyrolysis kinetics of ethylene-propylene (EPM) and ethylene-propylene-diene (EPDM). *Polymer Degradation and Stability*, **98**((9), 1571-1577.

47. Madorasky, S. L., and Strauss, S. (1961) *Thermal Degradation of Polymers*, Society of Chemical Industry, Monograph number 13.

48. Wall, L. A., and Strauss, S. (1960) Pyrolysis of polyolefins. *Journal of Polymer Science*, **44**, 313 -323.

49. Smith, W. K., and King, J. B. (1970) Surface temperature of materials during radiant heating to ignition. *Journal of Fire & Flammability*, **1**, 272-289.

50. Strauss, S., and Madorasky, S. L. (1962) Pyrolysis of some polyvinyl polymers at temperatures up to 1,200 °C. *Journal of Research of the National Bureau of Standards - A. Physics and Chemistry*, **66A**, 401-406.

51. Kutty, S. K. N., and Nando, G. B. (1993) Studies on the flammability of short Kevlar@ fibre-thermoplastic polyurethane composites. *Journal of Fire Sciences*, **11**(1), 66-79.

52. Gupta, Y. N., Chakraborty, A., Pandey, G. D., and Setua, D. K. (2003) High resolution thermo-gravimetry coupled mass spectrometry analysis of flame-retardant rubber. *Journal of Applied Polymer Science*, **89**, 2051-2057.

53. Setua, D. K., Pandey, G. D., Indushekhar, R., and Mathur, G. N. (2000) Biodeterioration of coated nylon fabric. *Journal of Applied Polymer Science*, **75**, 685-691.

54. Price D., Anthony, G. and Carty, P. (2001) Introduction: polymer combustion, condensed phase pyrolysis and smoke formation. In: *Fire Retardant Materials*, Horrocks, A. R., and Price, D. (eds.), Woodhead Publishing, UK, pp. 1-30.

55. Camino, G., Luda, M., and Costa, L. (1993) Combustion and fire retardance in polymeric materials. *Journal de Physique IV Colloque*, **03**(C7), 1539-1542.

56. Hilado, C. J. (1973) An overview of the fire behavior of polymers. *Fire Technology*, **9**(3), 198-208.

57. Mouritz, A. P., Mathys, Z., and Gibson, A. G. (2006) Heat release of polymer composites in fire. *Composites, Part A*, **37**, 1040-1054.

58. Albdiry, M. T., Almosawi, A. I., and Yousif, B. F. (2012) The synergistic effect of hybrid flame retardants on pyrolysis behaviour of hybrid composite materials. *Journal of Engineering Science and Technology*, **7**(3), 351-359.

59. Salmera, K. A., Gaan, S., and Malucelli, G. (2016) Recent advances for flame retardancy of textiles based on phosphorus chemistry. *Polymers*, **8**(9), 319.

60. Mansurov, Z. A., Ya. Kolesnikov, B., and Efremov, V. L. (2018) The role of carbonized layers for fire protection of polymer materials. *Eurasian Chemico Technological Journal*, **20**, 63-72.

61. Petrova, N. P., Ushmarin, N. F., Gnezdilov, D. O., and Koltsov, N. I. (2015) The development of a flame-retardant rubber compound based on general-purpose rubbers. *Kauchuk i Rezina*, **4**, 16-18.

62. Frasca, D., Schulze, D., Bohning, M., Krafft, B., and Schartel, B. (2016) Multilayer graphene chlorine isobutyl isoprene rubber nanocomposites: Influence of the multilayer graphene concentration on physical and flame retardant properties. *Rubber Chemistry & Technology*, **89**(2), 316-334.

63. Khattab, M. A., Feteha, F. A. H., Sadik, W. A., and Abdel-Bary, E. M. (2017) Effect of aluminium trihydrate as flame retardant on properties of a thermoplastic rubber nanocomposite. *Fire and Materials: An International Journal*, **41**(6), 688-699.

64. Zhang, C., and Wang, J. (2017) Natural rubber/dendrimer modified montmorillonite nanocomposites: Mechanical and flame-retardant properties. *Materials*, **11**(1), E41.

65. Gilman, J. W., Jackson, C. L., Morgan, A. B., Harris, R., Manias, E., Giannelis, E. P., Wuthenow, M., Hilton, D., and Phillips, S. H. (2000) Flammability properties of polymer-layered-silicate nanocomposites. Polypropylene and polystyrene nanocomposites. *Chemistry of Materials*, **12**, 1866-1873.

66. Zhu, J., Uhl, F. M., Morgan, A. B., and Wilkie, C. A. (2001) Studies on the mechanism by which the formation of nanocomposites enhances thermal stability. *Chemistry of Materials*, **13**(12), 4649-4654.

67. Dasari, A., Yu, Z. Z., Mai, Y. W., and Liu, S. (2007) Flame retardancy of highly filled polyamide 6/clay nanocomposites. *Nanotechnology*, **18**, 445602.

68. Samyn, F., Bourbigot, S., Jama, C., Nazare, S., Biswas, B., Hull, R., Fina, A., Castrovinci, A., Camino, G., Hagen, M., and Delichatsios, M. (2008) Characterisation of the dispersion in polymer flame retarded nanocomposite. *European Polymer Journal*, **44**, 1631-1641.

69. Laoutid, F., Bonnaud, L., Alexandre, M., Lopez-Cuesta, J. M., Dubois, P. H. (2009) New prospects in flame retardant polymer materials: From fundamentals to nanocomposites. *Materials Science and Engineering R: Reports*, **63**(3), 100-125.

70. Wang, L., He, X., and Wilkie, C. A. (2010) The utility of nanocomposites in fire retardancy. *Materials*, **3**, 4580-4606.

71. HaiYun, M., PingAn, S., and ZhengPing, F. (2011) Flame retarded polymer nanocomposites: Development, trend and future perspec-

tive. *Science China Chemistry*, **54**(2), 302-313.

72. Dehaghani, H. E., Ashori, D., and Soureshjani, M. H. (2013) Flame retarded Nylon 66 nanocomposites: Comparing the effect of different flame retardants on the flammability. *Asian Journal of Chemistry*, **25**(13), 7153-7157.

73. Mark, H. F., Atlas, S. M., Shalaby, S. W., and Pearce, E. M. (1975) Combustion of polymers and its retardation. In: *Flame Retardant Polymeric Materials*, Lewin, M., Atlas, S. M., and Pearce, E. M. (eds.), Springer, USA.

74. *Flame Retardant Phosphorus-containing Polymers*, Max Planck Institute for Polymer Research, Germany. Online: http://www.mpip-mainz.mpg.de/5133310/flameretardants [accessed 17th September 2018] (2018).

75. Beyer, G. (2003) Carbon Nanotube as a New Class of Flame Retardants for Polymers. Proceedings of the *52nd International Wire & Cable Symposium*, USA, pp. 628-633.

76. Plesa, I., Notingher, P. V., Schlogl, S., Sumereder, C., and Muhr, M. (2016) Properties of polymer composites used in high-voltage applications. *Polymers*, **8**(5), 173.

77. Hu, Y., Qian, X., Song, L., and Lu, H. (2014) Polymer/Layered Compound Nanocomposites: a Way to Improve Fire Safety of Polymeric Materials, Fire Safety Science. Proceedings of the *Eleventh International Symposium of International Association for Fire Safety Science*, New Zealand, pp. 66-82.

78. Lyon, R. E., and Janssens, M. L. (2015) Polymer flammability. In: *Encyclopedia of Polymer Science and Technology*, Wiley, USA.

79. Nazare, S., and Horrocks, A. R. (2008) Flammability testing of fabrics. In: *Fabric Testing*, Hu, J. (ed.), Woodhead Publishing, UK, pp. 339-388.

80. *Elements of Polymer Fire Safety and Guide to the Designer*, volume 5, The National Academies Press, USA (1979).

81. Zhang, F., Zhang, J., and Wang, Y. (2007) Modeling study on the combustion of intumescent fire-retardant polypropylene. *eXPRESS Polymer Letters*, **1**(3), 157-165.

82. Bal, N., and Rein, G. (2015) On the effect of inverse modelling and compensation effects in computational pyrolysis for fire scenarios. *Fire Safety Journal*, **72**, 68-76.

83. Vermesi, I., Roenner, N., Pironi, P., Hadden, R. M., and Rein, G. (2016) Pyrolysis and ignition of a polymer by transient irradiation. *Combustion and Flame*, **163**, 31-41.

7

Effect of Amine Functionalized Reduced Graphene Oxide on the Thermal, Mechanical, Rheological and Electrical Properties of CPE Compatibilized HDPE

7.1 Introduction

Polymer nanocomposites, in which one dimension of the filler is <100 nm, exhibit significant improvements in mechanical, gas barrier, thermal and fire retardancy properties, among others [1]. In recent years, graphene based polymer nanocomposites have attracted considerable research attention due to the potential of graphene in imparting extraordinary properties to the composites as compared to the pure polymer matrices [2-4]. Graphene layers are atomically thick two-dimensional (2-D) sheets composed of sp^2 carbon atoms arranged in a hexagonal lattice with large aspect ratio. Graphene possesses Young's modulus of 1 TPa, ultimate strength of 130 GPa, thermal conductivity of 5000 W/(m.K) and electrical conductivity up to 6000 S/cm, thus, underlining the role as a superior nanofiller in polymer matrices as compared to conventional fillers [4,5-8]. However, lack of polymer-graphene interfacial bonding, which originates due to the atomically smooth and non-reactive surface of graphene, inhibits the load transfer from the matrix to the graphene sheets across the polymer-filler interface. In addition, the graphene sheets tend to aggregate in the polymer matrices, thus, resulting in poor dispersion [9-11]. On the other hand, the dispersion of polar graphene oxide in nonpolar polymers like polyethylene also suffers due to the absence of positive interactions between them. In order to overcome this problem, compatibilizers have been commonly used to enhance the interfacial adhesion between the filler and host polymer, thus, resulting in improved filler dispersion [12,13]. Moreover, the functionalized graphene sheets provide multiple bonding sites with the polymer matrix or compatibilizer. Functionalization of graphene oxide (GO) sheets is

M. R. Vengatesan, H. Y. Al Asafen, A. M. Varghese and V. Mittal, The Petroleum Institute (part of Khalifa University of Science and Technology), Abu Dhabi, UAE*
Current address: Bletchington, Wellington County, Australia
© 2019 Central West Publishing, Australia

even much easier as compared to pristine graphene due to hydroxyl and epoxide functional groups on their basal planes, in addition to carbonyl and carboxyl groups located at the sheet edges [14]. Kim *et al.* [15] also reported the importance of functionalized polyethylene as well as blending methods for exfoliation of graphene in the poly-ethylene matrix. Similar work by Ren *et al.* [16] also reported high density polyethylene (HDPE) nanocomposites using functionalized graphene.

In the present study, the effect of amine functionalized reduced graphene oxide (r-AGO) on the crystallization, thermal stability and mechanical properties of HDPE has been studied, using chlorinated polyethylene (CPE) as compatibilizer. The objective was to develop HDPE/graphene composite materials with enhanced mechanical, thermal, rheological and electrical properties by generating covalent interactions between the compatibilizer and functionalized filler to enhance the degree of filler exfoliation.

7.2 Experimental

7.2.1 Materials

High-density polyethylene BB2581 was kindly supplied by Abu Dhabi Polymers Company Limited (Borouge), UAE. It had a melt flow index of 0.35 g/10 min (190 °C, 2.16 kg). Its melt temperature ranged be-tween 170-200 °C. Chlorinated polyethylene (CPE) Weipren 6025 (25% chlorine content, denoted as CPE25) was obtained from Lianda Corporation, USA. The CPE was in the form of white granules (specific gravity = 1.1-1.3). The melt flow index (190 °C, 2.16 kg) for CPE25 was 1.8 g/10 min and the heat of fusion was 45 J/g [17]. Graphene oxide was synthesized in the lab using Tour's method [18]. 1-(3-Di-methylaminopropyl)-3-ethylcarbodiimide hydrochloride (EDC.HCl), ethylene diamine (EA), dimethylformamide (DMF) and acetone were procured from Sigma Aldrich and were used without further purifi-cation.

7.2.2 Preparation of Amine Functionalized Reduced Graphene Oxide

r-AGO was prepared by following the reported literature with slight modification [8]. In brief, 900 mg of GO was dispersed by sonication and ultrasonic probe in 450 mL of DMF for 1 h at room temperature.

10 mL of EA and 200 mg of EDC.HCl were subsequently added to the suspension. The mixture was stirred at 110 °C for 24 h. The product, r-AGO, was diluted with 1000 mL of ethanol and left to settle for 24 h. Later, the superannuate was removed, and the product was filtered and washed with acetone to ensure the removal of any excess diamine. Finally, the products were dried at 50 °C overnight under vacuum.

7.2.3 Nanocomposite Preparation

Melt mixing was used to prepare nanocomposites of r-AGO with CPE25 and HDPE using mini twin conical screw extruder (MiniLab HAAKE Rheomex CTW5, Germany). A mixing temperature of 180 °C for 15 min at 100 rpm with batch size of 5 g was used. The screw length and screw diameter were 109.5 and 5/14 mm conical, respectively. The amount of CPE in the nanocomposites was fixed at 5.0 wt%, whereas r-AGO content was varied as 0.5, 1.0, 1.5, 3.0 and 5.0 wt%). Scheme 7.1 also represents the schematic of nanocomposite preparation. After extrusion, the samples were processed using injection molding (Thermo Scientific) at 190 °C with 400 bar injection pressure for 10 s in order to generate test specimens for mechanical and rheological analysis. In order to confirm the covalent bonding between the r-AGO and CPE, a hybrid of r-AGO/CPE was generated. In brief, 100 mg of r-AGO was dispersed in 100 mL of *o*-xylene using an ultra-sonication bath at 30 °C for 1 h. Subsequently, 100 mg of CPE was added to the dispersion and the mixture was stirred for 2 h at 130 °C. It was followed by the evaporation of the solvent under vacuum at 100 °C, and the resultant product was termed as r-AGO/CPE hybrid. For comparison purposes, HDPE/CPE blend was also generated by melt mixing, using the same processing conditions mentioned above.

7.2.4 Characterization Techniques

Fourier transform infrared (FT-IR) spectra were collected on Nicolet iS10 spectrometer equipped with SmartiTR diamond ATR accessory (angle of incidence of 45°), DTGS KBr detector and KBr beam splitter. It had a diamond ATR crystal (index of refraction = 2.4 at 1000 cm^{-1}) and a depth penetration of 2 μm at 1000 cm^{-1} for sample with refractive index of 1.5. Spectra were recorded by the OMNIC software in the 4000-400 cm^{-1} region with a resolution of 4 cm^{-1} from 32 scans. For

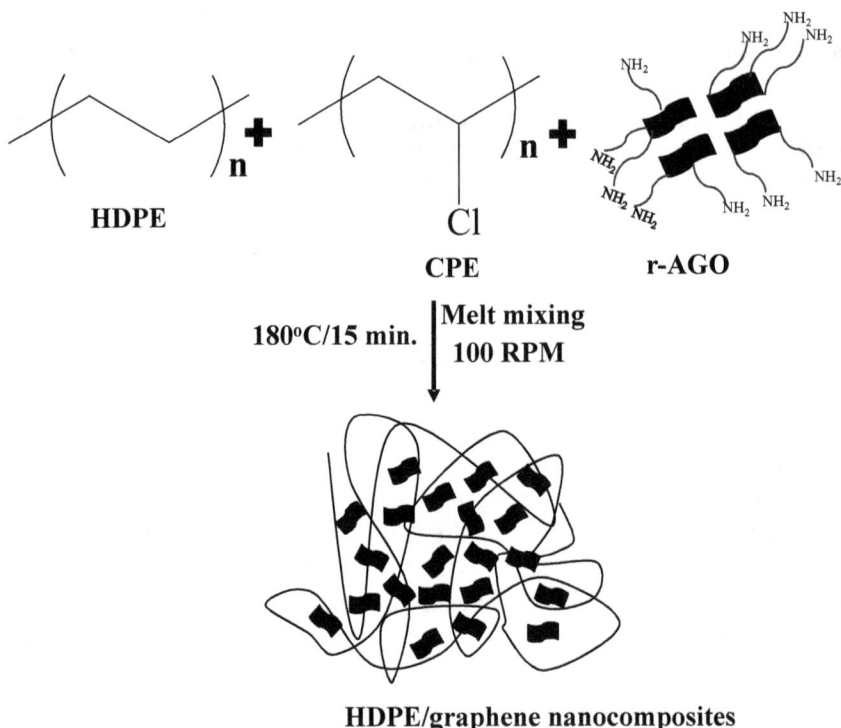

HDPE/graphene nanocomposites

Scheme 7.1 Schematic representation of generation of HDPE/graphene nanocomposites.

the microscopy analysis, ultrathin sections (30-70 nm) of the composite samples were obtained using a PowerTome equipped with a diamond knife at -30 °C. The sections were collected on 400 mesh formvar electron microscopy grids (coated with copper), and were examined in a FEI electron microscope (TECNAI) at 200 kV at room temperature without staining. Wide-angle X-ray diffraction (WAXRD) analysis of the filler and composites was performed using analytical powder diffractometer (X'Pert PRO) using Cu Kα radiation (λ = 1.5406 Å) in reflection mode. A zero-background holder was used to minimize the noise. The samples were step-scanned from 2θ = 5-60° at room temperature using a step size of 2θ = 0.02° and a step time of 10 s.

Discovery TGA from TA Instruments was used for the thermogravimetric analysis (TGA) of the samples. The test samples were heated from 35 to 700 °C under nitrogen atmosphere with a heating rate of 10 °C/min. Differential scanning calorimetric (DSC) analysis of the composites was performed on a TA Discovery DSC under nitrogen atmosphere. The scans were obtained from 35-200 °C using a heating rate of 10 °C/min. The second heating runs were carried out in the same temperature range and were used for calculation of the crystallinity [19]. 4-7 mg of the samples were used for the both DSC and TGA analysis. The mechanical testing of the nanocomposites was performed on universal testing machine (Instron) (ASTM D 638). The dumbbell shaped samples were used, and an average of five samples was recorded. A loading rate of 10 mm/min was employed, and the tests were carried out at room temperature. Bluehill Tree Analysis software was used for the calculation of tensile modulus, ultimate tensile strength (UTS) and tensile strain. The PRS-801 Resistance system was used to measure the surface resistance of the samples. It is capable of getting precision resistance measurements from 0.1 up to 2.0×10^{14} ohms with an overall measurement tolerance of ± 5%. Geometric averages of the resistance measured from 3 to 6 different spots from each side of the sample were recorded.

The rheological properties of the nanocomposites were measured by AR 2000 rheometer from TA Instruments. The disc shape samples of 25 mm diameter and 2 mm thickness was used to measure at 180 °C using a gap opening of 1.2 mm. Strain sweep scans were recorded at $\omega = 1$ rad/s from 0.1 to 100% strain and the samples were observed to be shear stable up to 10% strain. Frequency sweep scans of the samples were recorded at 4% strain from $\omega = 0.1$ to 500 rad/s

7.3 Results and Discussion

The FTIR spectra of CPE and CPE/r-AGO hybrid are presented in Figure 7.1. In the CPE spectrum, the peak at 668 cm^{-1} corresponded to the C-Cl stretching frequency. In addition, the peaks at 2926 and 2853 cm^{-1} corresponded to asymmetric and symmetric characteristic –CH$_2$ stretching vibrations of CPE. For the hybrid, a reduction in the intensity of the peak at 668 cm^{-1} for C-Cl stretching frequency was observed, which indicated the reaction between r-AGO and CPE25. Furthermore, the peak at 1029 cm^{-1}, corresponding to –CNO- stretching frequency, also represented the nucleophilic reaction between the amine groups in r-AGO and Cl group in CPE [18].

The calorimetric values are also depicted in Table 7.1. The heat of fusion of pure crystalline HDPE was taken as 293 J/g and was used to determine the percentage crystallinity in the polymer nanocomposites [8]. The crystallinity of the samples decreased from 59% to 54%

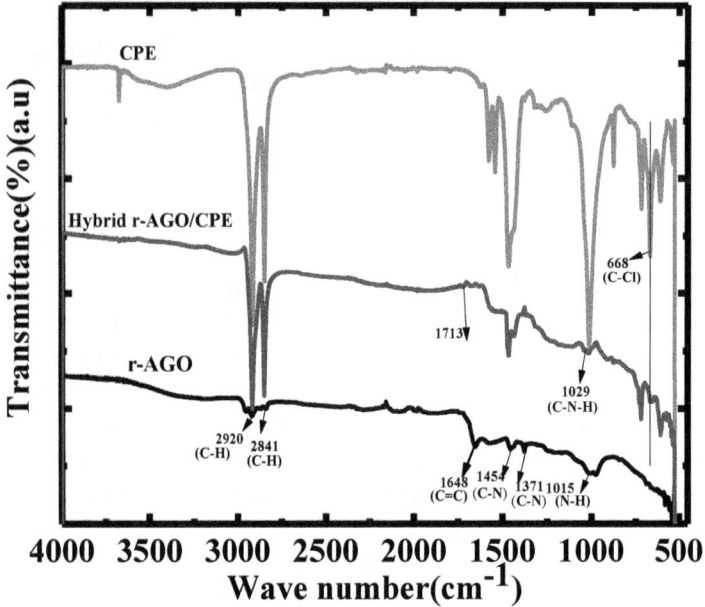

Figure 7.1 FT-IR spectra of r-AGO, r-AGO/CPE and CPE.

Table 7.1 Calorimetric properties of HDPE/r-AGO nanocomposites

Sample	ΔH (J/g)	T_m (°C)	Crystallinity (%)	T_c (°C)
Pure HDPE	174	135	59	119
HDPE/5.0 wt% CPE	158	135	54	122
HDPE/CPE/0.5 wt% r-AGO	174	133	60	120
HDPE/CPE/1.0 wt% r-AGO	175	135	60	120
HDPE/CPE/1.5 wt% r-AGO	175	135	61	119
HDPE/CPE/3.0 wt% r-AGO	174	135	61	119
HDPE/CPE/5.0 wt% r-AGO	161	136	58	119

on the addition of 5 wt% CPE25. This indicated that the CPE25 component did not act as a nucleating agent for the polyethylene chains. The melt enthalpy increased in the nanocomposites with r-AGO content till 3 wt% filler content. For the composite with 5 wt% r-AGO, the melt enthalpy decreased to a value of 161 J/g as compared to 174 J/g for pure HDPE. Thus, the degree of crystallinity increased with increase in r-AGO content up to 3 wt%, which indicated that the r-AGO acted as a nucleating agent for the polyethylene chains. However, a slight reduction in the degree of crystallinity for the 5 wt% r-AGO composite indicated that at high filler concentration, r-AGO tended to aggregate in HDPE and resulted in a decrease in crystallinity. Thermal degradation temperature for 10 and 40% weight loss as well as char yield value at 700 °C are also presented in Table 7.2. From the table, it was observed that the addition of r-AGO resulted in almost similar degradation behavior as HDPE/CPE blend, but was significantly improved as compared to pure HDPE. The char yield of the nanocomposites increased with an increase in the r-AGO content. For example, the char yield of 5 wt% r-AGO filled HDPE was 3.2% as compared to 0% for the HDPE/CPE blend at 700 °C. This indicated that r-AGO acted as thermal barrier and reduced the thermal degradation of the polymer at higher temperature [20].

Table 7.2 Thermal stability and char yield of HDPE/r-AGO nanocomposites

Sample	$T_d{}^{10}$ (°C)	$T_d{}^{40}$ (°C)	Char yield at 700 °C (%)
HDPE	428	462	0.00
HDPE/5 wt% CPE	447	473	0.00
HDPE/CPE/0.5 wt% r-AGO	447	473	0.37
HDPE/CPE/1.0 wt% r-AGO	447	476	1.25
HDPE/CPE/1.5 wt% r-AGO	450	476	1.26
HDPE/CPE/3.0 wt% r-AGO	450	476	2.30
HDPE/CPE/5.0 wt% r-AGO	451	478	3.20

$T_d{}^{10}$ - 10% weight loss temperature
$T_d{}^{40}$ - 40% weight loss temperature

As presented in Table 7.3, the tensile modulus of pure HDPE was observed to be 1033 MPa. Addition of 5 wt% CPE reduced the modulus of HDPE by about 8% probably due to matrix plasticization by the lower molecular weight CPE chains. An enhancement of the tensile modulus was realized with increase in content of r-AGO up to 3 wt%. Further increasing the content of r-AGO (5 wt%) resulted in a decrease in the modulus value probably due to filler aggregation. Overall, the observed results indicated that the functional groups on the graphene surface effectively enhanced the interfacial adhesion between the graphene and polymer matrix, due to the covalent bonding of the ethylene diamine modified GO (r-AGO) with the HDPE/CPE blend *via* nucleophilic reaction, thus, resulting in improved composite properties. The observation is also in good agreement with the reported literature [21]. Therefore, the strong interfacial adhesion between the HDPE and r-AGO led to improved interfacial stress transfer mechanism and restricted mobility of polymer chains in the nanocomposites. The addition of r-AGO didn't negatively impact the tensile strength of the nanocomposites till 3 wt%, however, the addition of 5 wt% of r-AGO reduced the tensile strength from 29 MPa to 22 MPa. As mentioned earlier, higher loading of r-AGO may form aggregates in the matrix interface and decrease the interfacial adhesion. This is in good agreement with expanded graphite-high density polyethylene nanocomposites reported earlier [23].

Table 7.3 Mechanical properties of HDPE/r-AGO nanocomposites

Sample	Tensile strength (MPa)	Young's modulus (MPa)
Pure HDPE	28.3 (±5)	1003 (±10)
HDPE/5 wt% CPE	29.2 (±5)	949 (±10)
HDPE/CPE/0.5 wt% r-AGO	27.8 (±5)	1009 (±10)
HDPE/CPE/1.0 wt% r-AGO	29.1 (±5)	1026 (±10)
HDPE/CPE/1.5 wt% r-AGO	28.6 (±5)	1051 (±10)
HDPE/CPE/3.0 wt% r-AGO	27.6 (±5)	1122 (±10)
HDPE/CPE/5.0 wt% r-AGO	24.2 (±5)	1005 (±10)

The rheological behavior of the HDPE/r-AGO nanocomposites was studied to understand the network structure formation, dispersion level of r-AGO within the HDPE matrix and interfacial interactions between the r-AGO and HDPE matrix [24,25]. The storage (G') and loss moduli (G") of the HDPE matrix and nanocomposites are presented in Figure 7.2(a-b). Addition of 5 wt% of CPE to HDPE resulted in a decrease in the values of both G' and G", as the addition of CPE enhanced the shear behavior of HDPE during molecular mixing [26]. G' and G" values increased with increase in the content of r-AGO up to 1 wt% and further increase in the r-AGO content resulted in a gradual decrease. Addition of 5 wt% of r-AGO exhibited a significant decrease in the storage and loss modulus values. A well-defined plateau of G' was observed for the HDPE nanocomposites at lower frequency which indicating a transition from the liquid-like to pseudo solid-like viscoelastic behavior, thus, leading to percolation network structure [27,28]. Decrease in G' of 5 wt% r-AGO nanocomposites as compared to the pure HDPE was probably due to the sheet stacking of the r-AGO phase at higher loading. Figure 7.2(c) shows the complex viscosity (η^*) as a function of angular frequency. The horizontal curves of the η^* values of the HDPE and nanocomposites represented a Newtonian fluid, and the η^* values decreased with an increase in angular frequency owing to shear thinning [29]. The η^* of the nanocomposites increased up to 1.0 wt% of r-AGO content. Similar to the modulus, further increasing the content above 1.0 wt% resulted in a gradual decrease of η^*. This can be attributed to the selective physico-absorption of polymer chains on the r-AGO surface, along with de-entanglement effect [30]. The addition of 5 wt% r-AGO exhibited a significant decrease in η^*. As mentioned earlier, the higher r-AGO loading led to sheet stacking due to hydrodynamic effect, and the applied shear had a lesser effect on the re-entanglement due to higher extent of agglomeration of the r-AGO phase [31,32].

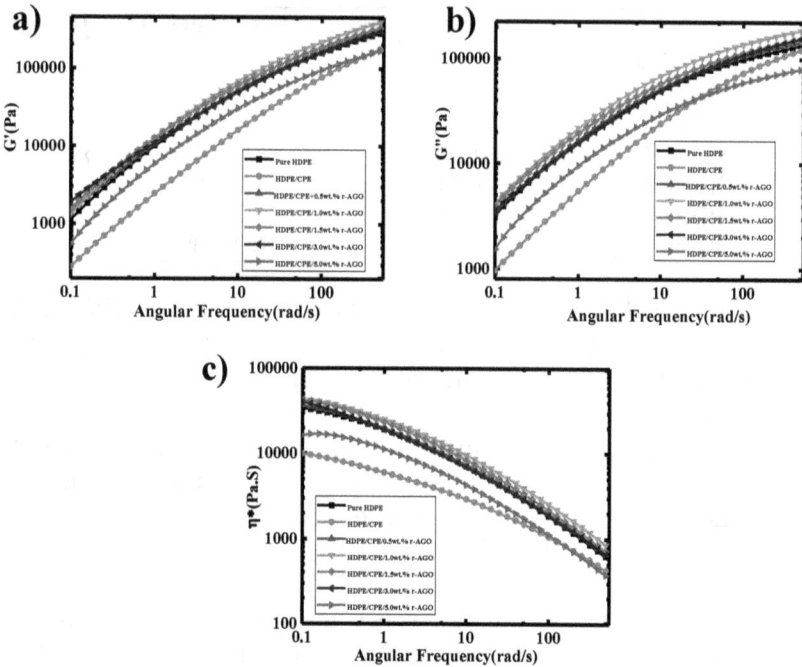

Figure 7.2 (a) Storage modulus (G'), (b) loss modulus (G") and (c) complex viscosity (η^*) of HDPE/r-AGO nanocomposites.

The incorporation of the conducting nanofillers in the polymer matrix leads to the generation of conducting pathways for electrical conduction in the polymer nanocomposites *via* percolation of filler [15]. To analyze such an effect, surface resistivity of the samples was measured and the values are presented in Table 7.4. Pure HDPE and HDPE/CPE had high surface resistivity due to the lack of any conducting pathways [33]. Addition of r-AGO decreased the surface resistivity of the nanocomposites up to 3.0 wt%, indicating that the composites had enhanced conductivity. In fact, the resistance decreased by two orders of magnitude on the addition of 3.0 wt% of filler. Further increase in the content of r-AGO did not enhance the conductivity. The rapid transition and/or unchanged conductivity represented the achievement of percolation threshold [34]. This indicated that 3 wt% filler fraction was suitable to form a conducting network in the HDPE matrix.

Table 7.4 Surface resistivity of HDPE/r-AGO nanocomposites

Sample	Surface Resistance (Ohm)
Pure HDPE	14×10^{10}
HDPE + 5.0 wt% CPE	15×10^{10}
HDPE / CPE/ 0.5 wt% r-AGO	12×10^{9}
HDPE / CPE/ 1.0 wt% r-AGO	9.3×10^{9}
HDPE/CPE/1.5 wt% r-AGO	3.4×10^{9}
HDPE/CPE/3.0 wt% r-AGO	11.7×10^{8}
HDPE/CPE/5.0 wt% r-AGO	10.3×10^{9}

In addition, Table 7.5 shows the surface resistivity of HDPE/CPE/3 wt% r-AGO nanocomposites with other graphene based polymer nanocomposites reported in literature [4,21,35-37]. The conductivity of the nanocomposites varied with the nature and amount of filler, thickness of the composite film and applied voltage. The conductivity of the graphene nanocomposites mainly depended on the nature of graphene nanofiller such as reduced graphene oxide, expanded graphite, thermally reduced graphene oxide, functionalized graphene oxide and graphene oxide.

Figure 7.3(a) shows the XRD spectra of pure GO and r-AGO. Pure GO exhibited an intense peak at $2\theta = 10.0°$ corresponding to (002) with *d*-spacing of 8.8 Å. After reduction of GO using EDA, the intensity of peak at 10.0° reduced and a broad peak appeared at $2\theta = 26.5°$ was observed. This indicated that the majority of the oxide and water molecules were removed and the EDA intercalated in between the GO layers [35]. Figure 7.3(b) shows the diffraction patterns of HDPE, HDPE/CPE blend and HDPE/r-AGO nanocomposites. The peaks at 2θ

Table 7.5 Comparison of the electrical properties of HDPE/CPE/r-AGO nanocomposites with other polymer/graphene nanocomposites reported in literature

Polymer nano-composites	Graphene	Filler content (wt%)	Surface resistivity (Ω)/electrical conductivity (S/m or S/cm)	Ref.
Graphene/poly-ethylene nano-composites	Thermally reduced graphene oxide	3	4.0×10^6 Ω	[4]
Hexamethylene diamine function-alized graphene oxide/polypropyl-ene nanocomposites	hexamethylene diamine functional-ized graphene oxide	2.0	1×10^5 Ω	[21]
Polyethylene/reduced graphite oxide nanocomposites	Reduced graphite oxide	3.1	1.1×10^5 S cm^{-1}	[35]
Graphene based unsaturated poly-ester resin composites	Graphene nano-sheets	0.1	1×10^{15} Ω	[36]
Poly(methyl methacrylate) (PMMA)-reduced graphene oxide composite	Reduced graphene oxide	2.7	1×10^2 S m^{-1}	[37]
HDPE/CPE/r-AGO	Ethylene diamine functionalized reduced graphene oxide(r-AGO)	3.0	11.7×10^6 Ω	Present work

= 22.0° (110) and 24.3° (200) were attributed to characteristic diffraction peaks of pure polyethylene. From the diffraction patterns of the nanocomposites, it was observed that the addition r-AGO did not

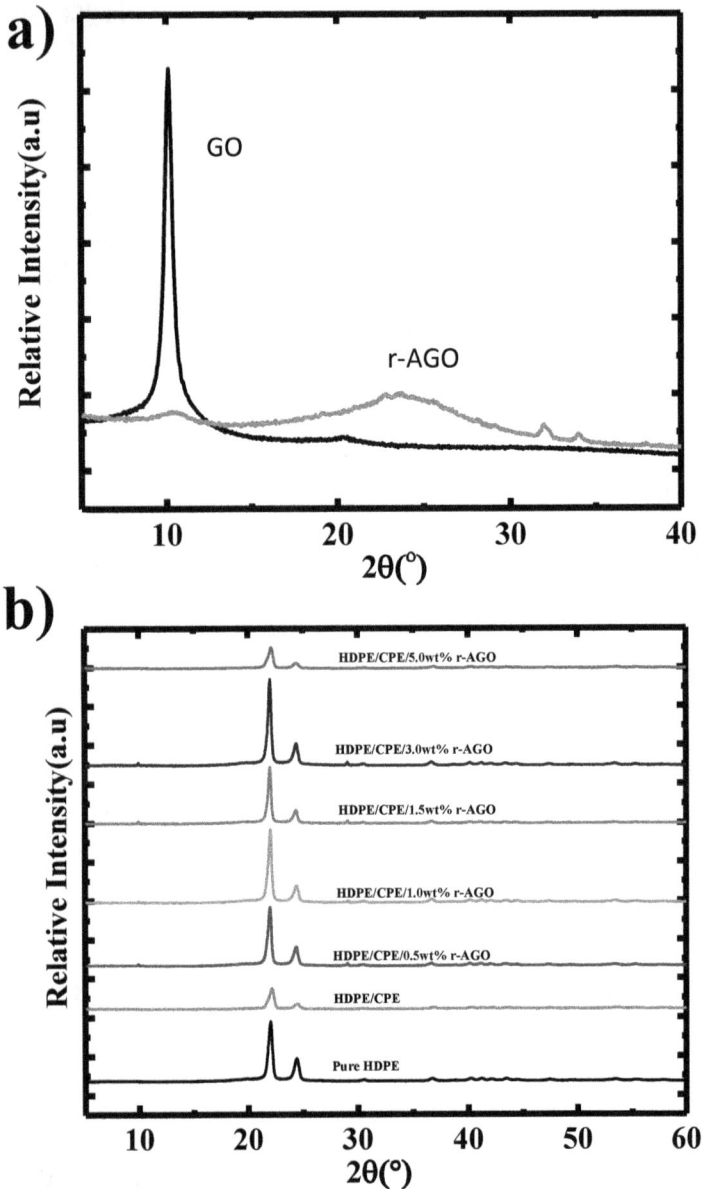

Figure 7.3 XRD spectra of (a) GO and r-AGO; (b) HDPE/r-AGO nanocomposites.

affect the crystal structure of HDPE. In addition, the peak at 26.5° was not observed in the diffraction patterns of HDPE/r-AGO nanocomposites. It indicated that r-AGO was exfoliated in the HDPE matrix, thus, also confirming the earlier findings. In addition, the peak intensity also increased with addition of r-AGO up to 3 wt% qualitatively indicating the increase in percentage of crystallinity. This suggested that the r-AGO acted as a nucleating agent in HDPE nanocomposites, which also confirmed the earlier DSC findings. Figure 7.4 also shows the TEM images of the nanocomposites. It was observed that the r-AGO was well dispersed in the polymer matrix till 3 wt% filler content. However in the case of 5 wt% AGO, aggregates were formed which resulted in poor dispersion in the polymer matrix.

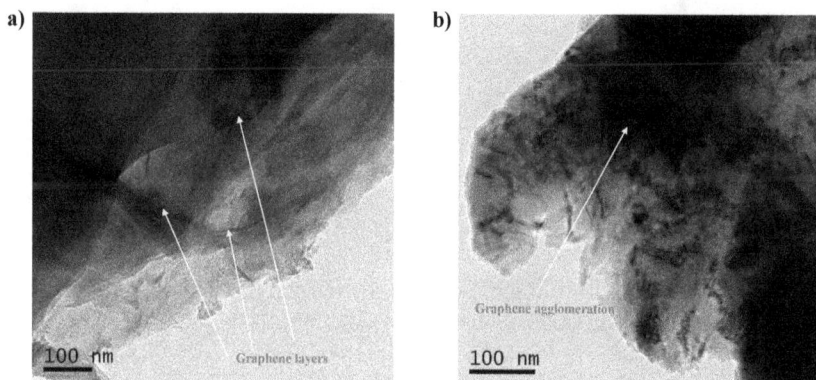

Figure 7.4 TEM images of composites with (a) 1.0 wt% and (b) 5.0 wt% filler content.

7.4 Conclusions

In this study, HDPE-graphene nanocomposites were developed using amine functionalized reduced graphene oxide (r-AGO) as nanofiller and CPE as compatibilizer. r-AGO covalently interacted with CPE *via* nucleophilic reaction, thus, enhancing the filler dispersion in the polymer. Owing to the improved interfacial interactions, enhancement in the thermal, mechanical and electrical properties of the resulting nanocomposites was observed. The Young's modulus of the composites increased marginally up to 3 wt% r-AGO content. The surface resistivity of the nanocomposites decreased from 14×10^{10} Ω to 11.7×10^8 Ω with an increase in the content of r-AGO up to 3.0 wt%. The storage modulus and complex viscosity increased up to 1 wt% filler content and gradually reduced with further increase in the content of r-AGO, though significant reduction was observed for 5 wt%

composite. Addition of r-AGO did not impact the crystalline structure of the polymer. TEM analysis exhibited uniform dispersion of filler in the composites up to 3 wt% content. The developed nanocomposites have the potential for applications in the field of microelectronics and packaging industries.

Authors' Contributions

M. R. Vengatesan and H. Y. Al Asafen contributed equally to this work.

Acknowledgements

The authors sincerely thank the PI Research Centre, The Petroleum Institute for the financial support.

References

1. Koo, J. H. (2006) *Polymer Nanocomposites: Processing, Characterization, And Applications*, Mcgraw-Hill, Nanoscience and Technology Series, USA.
2. Stankovich, S., Dikin, D. A., Dommett, G. H. B., Kohlhaas, K. M., Zimney, E. J., Stach, E. A., Piner, R. D., Nguyen, S. T., and Ruoff, R. S. (2006) Graphene-based composite materials. *Nature*, **442**, 282-286.
3. Cai, D. Y., and Song, M. (2010) Recent advance in functionalized graphene/polymer nanocomposites. *Journal of Materials Chemistry*, **20**, 7906-7915.
4. Kim, H., Abdala, A. A., and Macosko, C. W. (2010) Graphene/polymer nanocomposites. *Macromolecules*, **43**, 6515-6530.
5. Lee, C., Wei, X., Kysar, J. W., and Hone, J. (2008) Measurement of the elastic properties and intrinsic strength of monolayer graphene. *Science*, **321**, 385-388.
6. Balandin, A. A., Ghosh, S., Bao, W., Calizo, I., Teweldebrhan, D., Miao, F., and Lau, C. (2008) Superior thermal conductivity of single-layer graphene. *Nano Letters*, **8**, 902-907.
7. Tombros, N., Veligura, A., Junesch, J., Guimares, M., Veramarun, I., Jonkman, H., and Wees, B. (2011) Quantized conductance of a suspended graphene nanoconstriction. *Nature Physics*, **7**, 697-700.
8. Bunch, J., Verbridge, S., Alden, J., van der Zande, A., Parpia, J., Craighead, H., and McEuen, P. (2008) Impermeable atomic membranes from graphene sheets. *Nano Letters*, **8**, 2458-2462.
9. He, Q., Sudibya, H. G., Yin, Z., Wu, S., Li, H., Boey, F., Huang, W., Chen, P., and Zhang, H. (2010) Centimeter long and large scale micropat-

terns of reduced graphene oxide films: Fabrication and sensing applications. *ACS Nano*, **4**, 3201-3208.

10. Shen, J., Huang, W., Wu, L., Hu, Y., and Ye, M. (2007) Thermophysical properties of epoxy nanocomposites reinforced with amino functionalized multiwalled carbon nanotubes. *Composites Part A,* **38**, 1331-1336.

11. Wakabayashi, K., Pierre, C., Dikin, D. A., Ruoff, R. S., Ramanathan, T., Brinson, C., and Torkelson, J. M. (2008) Polymer graphite nanocomposites: Effective dispersion and major property enhancement via solid state shear pulverization. *Macromolecules*, **41**, 1905-1908.

12. Mittal, V., and Al Zaabi, K. (2013) Biodegradable polyester nanocomposites: Phase miscibility and properties. *Journal of Applied Polymer Science*, **130**(1), 516-525.

13. Schniepp, H. C., Li, J. L., McAllister, M. J., Sai, H., Herrera-Alonso, M., Adamson, D. H., Prudhomme, R. K., Car, R., Saville, D. A., and Aksay, I. A. (2006) Functionalized single graphene sheets derived from splitting graphite oxide. *Journal of Physics Chemistry B*, **110**, 8535-8539.

14. Chen, Y., Qi, Y., Tai, Z., Yan, X., Zhu, F., and Xue, Q. (2012) Preparation, mechanical properties and biocompatibility of graphene oxide/ ultrahigh molecular weight polyethylene composites. *European Polymer Journal*, **48**, 1026-1033.

15. Kim, H., Kobayashi, S., AbdurRahim, M. A., Zhang, M. J., Khusainova, A., Hillmyer, M. A., Abdala, A. A., and Macosko, C. W. (2011) Graphene/polyethylene nanocomposites: Effect of polyethylene functionalization and blending methods. *Polymer,* **52**, 1837-1846.

16. Ren, P., Wang, H., Huang, H., Yan, D., and Li, Z. (2014) Characterization and performance of dodecyl amine functionalized graphene oxide and dodecyl amine functionalized graphene/high-density polyethylene nanocomposites: A comparative study. *Journal of Applied Polymer Science*, **131**, doi: 10.1002/app.39803.

17. Mittal, V., Luckachan, G. E., and Matsko, N. B. (2013) PE/chlorinated-PE blends and PE/chlorinated-PE/graphene oxide nanocomposites: morphology, phase miscibility, and interfacial interactions. *Macromolecular Chemistry and Physics,* **215**, 255-268.

18. Marcano, D. C., Kosynkin, D. V., Berlin, J. M., Sinitskii, A., Sun, Z., Slesarev, A., Alemany, L. B., Lu, W., and Tour, J. M. (2010) Improved synthesis of graphene oxide. *ACS Nano*, **4**, 4806-4814.

19. Jiang, X., and Drzal, L. T. (2012) Multifunctional high-density polyethylene nanocomposites produced by incorporation of exfoliated graphene nano platelets 2: crystallization, thermal and electrical properties. *Polymer Composites*, **33**, 636-642.

20. Zhao, S., Chen, F., Zhao, C., Huang, Y., Dong, J. Y., and Han, C. C. (2013) Interpenetrating network formation in isotactic polypropylene/graphene composites. *Polymer*, **54**, 3680-3690.

21. Ryu, S. H., and Shanmugharaj, A.M. (2014) Influence of hexamethylene diamine functionalized graphene oxide on the melt crystallization and properties of polypropylene nanocomposites. *Materials Chemistry and Physics*, **146**, 478-486.

22. Achaby, M. E., Arrakhiz, F.-E., Vaudreuil, S., Qaiss, A. K., Bousmina, M., and Fassi-Fehri, O. (2012) Mechanical, thermal, and rheological properties of graphene-based polypropylene nanocomposites prepared by melt mixing. P*olymer Composites*, **33**, 733-744.

23. Lee, S. H., Cho, E. N., Jeon, S. H., and Youn, J. R. (2007) Rheological and electrical properties of polypropylene composites containing functionalized multi-walled carbon nanotubes and compatibilizers. *Carbon,* **45**, 2810-2822.

24. Potschke, P., Fornes, T. D., and Paul, D. R. (2002) Rheological behavior of multiwalled carbon nanotube/polycarbonate composites. *Polymer*, **43**, 3247-3255.

25. Chaudhry, A. U., and Mittal, V. (2013) High-density polyethylene nanocomposites using masterbatches of chlorinated polyethylene/graphene oxide. *Polymer Engineering and Science,* **53**, 78-88.

26. Achaby, M. E., and Qaiss, A. (2013) Processing and properties of polyethylene reinforced by graphene nanosheets and carbon nanotubes. *Materials and Design,* **44**, 81-89.

27. Horst, M. F., Quinzani, L. M., and Failla, M. D. (2014) Rheological and barrier properties of nanocomposites of HDPE and exfoliated montmorillonite. *Journal of Thermoplastic Composite Materials*, **27**, 106-125.

28. Li, Y., Zhu, J., Wei, S., Ryu, J., Sun, L., and Guo, Z. (2011) Poly (propylene)/graphene nanoplatelet nanocomposites: melt rheological behavior and thermal, electrical, and electronic properties. *Macromolecular Chemistry and Physics,* **212**, 1951-1959.

29. She, Y., Chen, G., and Wu, D. (2007) Fabrication of polyethylene/graphite nanocomposite from modified expanded graphite. *Polymer International*, **56**, 679-685.

30. Tang, Y., Gao, P., Ye, L., and Zhao, C. (2010) Organoclay-modified thermotropic liquid crystalline polymers as viscosity reduction agents for high molecular mass polyethylene. *Journal of Materials Science*, **45**, 5353-5363.

31. Tuteja, A., Mackay, M. E., Hawker, C. J., and Van Horn, B. (2005) Effect of ideal, organic nanoparticles on the flow properties of linear polymers: Non-Einstein like behavior. *Macromolecules*, **38**, 8000-8011.

32. Thomas, S. P., De, S. K., and Hussein, I. A. (2013) Impact of aspect ratio of carbon nanotubes on shear and extensional rheology of polyethylene nanocomposites. *Applied Rheology*, **23**, 23635.

33. Kaushik, A., Singh, P., and Bhagat, S. (2009) Preparation and characterization of graphite flakes-filled polyester composites. *Polymer-*

Plastics Technology and Engineering, **48**, 802-807.

34. Perez, L. D., Zuluaga, M. A., Kyu, T., Mark, J. E., and Lopez, B. L. (2009) Preparation, characterization, and physical properties of multiwall carbon nanotube/elastomer composites. *Polymer Engineering and Science*, **49**, 866-874.

35. Pavoski, G., Maraschin, T., Milani, M. A., Azambuja, D. S., Quijada, R., Moura, C. S., Basso, N. S., and Galland, G. B. (2015) Polyethylene/reduced graphite oxide nanocomposites with improved morphology and conductivity. *Polymer*, **81**, 79-86.

36. Swain, S. (2013) Synthesis and characterization of graphene based unsaturated polyester resin composites. *Transactions on Electrical and Electronic Materials*, **14**, 53-58.

37. Pham, V. H., Dang, T. T., Hur, S. H., Kim, E. J., and Chung, J. S. (2012) Highly conductive poly(methyl methacrylate) (PMMA)-reduced graphene oxide composite prepared by self-assembly of pmma latex and graphene oxide through electrostatic interaction. *ACS Applied Materials & Interfaces*, **4**, 2630-2636.

38. Navaee, A., and Salimi, A. (2015) Efficient amine functionalization of graphene oxide through the Bucherer reaction: an extraordinary metal-free electrocatalyst for the oxygen reduction reaction. *RSC Advances*, **5**, 59874-59880.

8

Liquid Crystalline Polymer Composites

8.1 Introduction

Properties of the liquid crystals (LCs) and polymers are synergistically combined in the liquid crystalline polymers (LCPs) [1,2]. Novel high performance polymeric materials have been developed using the anisotropy present in the liquid crystalline mesophase. The liquid crystalline polymers exhibit significant ability to form high strength fibers [3-6]. In addition, the aromatic liquid crystalline polymers have been used in many advanced products, formed using injection molding. Binary and ternary blends of these polymers have also been developed [7]. The magnetic properties of the liquid crystalline polymers consisting of lyotropic or thermotropic LCs have also been utilized for different uses. The isotropic LCs included DNA, polyribonucleotides and polypeptides, whereas the thermotropic LCs comprised polypeptides and polyesters. Numerous studies have reported the development of LCPs using different polymerization techniques [8-26]. The ease of processing and optimal properties also make them attractive materials for developing nanocomposites with fillers like titanium dioxide (TiO_2), silicates, montmorillonite (MMT), carbon nanotubes (CNTs), graphene, etc.

8.2 Liquid Crystals

The common characteristics of the LCs include polarizability, strong dipoles, rigidness of the long axis, etc. [27-28]. One of the most distinguishable property of the liquid crystals is the ability to direct towards a common axis, termed as director. The molecules in a liquid crystal are generally referred to as mesogens. The molecular order in the liquid crystals is different from the order of molecules in the liquid phase, where no particular order is demonstrated. On the other hand, in the solid state, there is no free molecular movement, and the

Liyamol Jacob and Vikas Mittal*, The Petroleum Institute (part of Khalifa University of Science and Technology), Abu Dhabi, UAE
*Current address: Bletchington, Wellington County, Australia
© 2019 Central West Publishing, Australia

molecules are highly ordered. The mesogenic phase in the liquid crystals, thus, exhibits a molecular arrangement in between the liquid and solid phases [29-31]. In terms of order parameters, the liquid crystals exhibit a very high order compared to the isotropic liquids. However, they still demonstrate a high degree of freedom in terms of molecular oscillation, translation, rotation and conformational changes, among others [32-35].

8.3 Classification of LCPs

The LCPs are mainly classified as the main chain LCPs and side chain LCPs, with differences in properties.

8.3.1 Main Chain LCPs

A flexible spacer helps in linking the mesogenic groups in the polymer scaffold, and the resulting materials are called the main chain LCPs. The mesophase structure is generally referred to as the low molecular weight structure. The polymer structure must be compatible enough to support the structure of the mesogenic phase. This results in a strong coupling between the mesogenic and polymeric properties. A large number of main chain LCPs like poly(azomethine-ester)s, poly(azomethine)s, poly(esterimide)s, polyimides, polyethers, polycarbonates, polyurethanes, poly(p-phenylene)s, polyamides and polyesters have been widely studied [36-54].

8.3.2 Side Chain LCPs

The side chain LCPs consist of mesogenic side chains, thus, avoiding the use of methylene spacers [55-62]. Success from the earlier studies specifically motivated the development of the side chains based on methacrylates and the backbones consisting of the acrylate/methacrylate chains [63-74]. An orientational order is achieved by packing the mesogens in a parallel order. Compared to the main chain LCPs, the orientation occurs faster in the side chain LCPs. Owing to this, the use of electric field leads to a significant and swift change in the optical properties of the side chain polymers.

At temperatures greater than the glass transition temperature, the anisotropic ordering of the main and side chains is prevented due to the exceptional mobility shown by the chain segments in the liquid phase. In case a flexible spacer has been used to connect the main chain and mesogenic groups, a decoupling in the motion of the main and side chains may take place. In such cases, the conformation in the main chain results in the inhibition of the anisotropic ordering even though the mesogenic constituents are able to develop a long range orientation order [75,76].

8.4 Liquid Crystalline Polymer Nanocomposites

Change in the molecular structure of the liquid crystalline polymers can be achieved by forming nanocomposites with a range of nanoparticles. [77-84]. Quantum size effects, increased surface area, enhancement in the active surface, etc., are some of the properties achieved by introducing nanoparticles to the LCP matrices [85,86].

Grafting to polymerization was used for functionalizing CdTe, ZnO, SnO_2 and TiO_2 nanorods with polymers such as poly(methylmethacrylate), polystyrene and poly(diethylene glycol monomethyl ether) methacrylate deblock copolymers containing anchor groups [87]. Reversible addition-fragmentation chain transfer (RAFT) polymerization was used to synthesize the block copolymers. The surface coverage of the nanorods was observed to be dependent on the block lengths and ratios, as determined from the thermogravimetric analysis. The maximum surface coverage was noted for the short anchor unit blocks. Stable dispersions formed from the hairy rod like structures exhibited liquid crystalline behavior in different solvents and in polymeric and oligomeric matrices. In another study, the side chain liquid crystalline siloxanes doped with magnetic cobalt nanoparticles were studied for their structural and orientation properties [88]. At room temperature, the materials exhibited a combination of both ferromagnetic properties and orientational behavior. The nanorods were observed to self-assemble in bundles consisting of rows packed in a 2-dimensional hexagonal lattice (Figure 8.1).

Zorn and Zentel [89] reported the use of RAFT copolymerization to synthesize narrowly distributed block copolymers consisting of a hole conducting triarylamine block and an anchor block. Reactive ester approach was used to introduce the anchor block. The dopamine anchor groups containing copolymers were observed to bind to different semiconductors like ZnO, SnO_2 and TiO_2. Stable dispersions in particular solvents were achieved for the poly(triphenylamine) grafted inorganic nanorods. The nanorods demonstrated liquid crystalline behavior at higher concentration in various solvents and an oligotriphenylamine matrix.

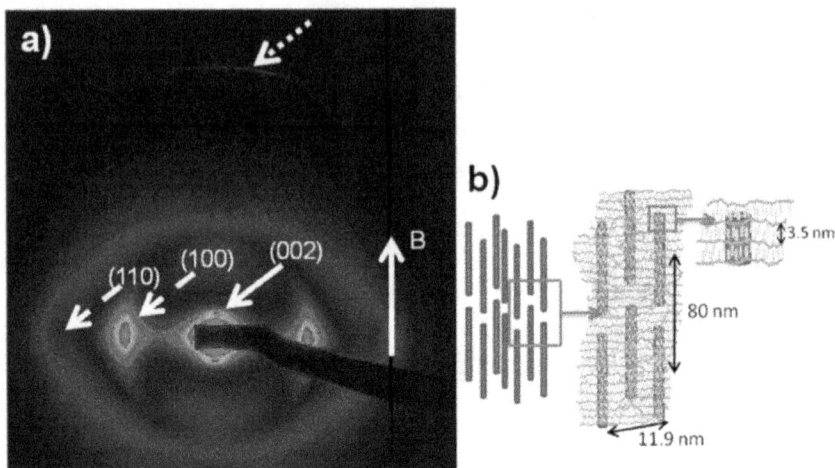

Figure 8.1 (a) Small angle X-ray scattering pattern of the Co nanorods hybrid material. The solid arrow points to the (001) reflection due to the organization of the Co nanorods in rows within the bundles whereas the dashed arrows point to the (100) and (110) reflections of the hexagonal 2-dimensional lattice and (b) schematic representation of the nanorod bundles. Reproduced from Reference 88 with permission from American Chemical Society.

Numerous other literature studies have also reported the development of new LCs and LCP composite materials [90-99]. Jang and Bae [93] reported the liquid crystalline epoxy/polyaniline (LCE/PANI) composite nanowires using an anodic aluminum oxide membrane. A temperature-gradient curing process was employed by the authors to develop the nanocomposites. Ezhov *et al.* [94] reported the nanocomposites of smectic C hydrogen-bonded polymers from the family of poly(4-(n-acryloyloxyalkoxy)benzoic acids with CdS nanorods (Figure 8.2). Uniaxial deformation of the composite films exhibited a long-range orientation with the formation of one-dimensional aggregates of rods. In another study, Horsch *et al.* [97] reported the self-assembly of nanorods functionalized by a polymer "tether." Meuer *et al.* [98] also functionalized TiO_2 nanorods with dopamine-functionalized diblock copolymers and observed the liquid crystalline behavior.

CNTs are attractive materials for reinforcing the host polymer matrices due to the properties like high aspect ratio, unique structural arrangement of atoms, outstanding mechanical, thermal and electrical performance, etc. [100-118]. The thermodynamic instability of thermotropic liquid crystalline polymers makes them immiscible with most thermoplastics at molecular level. The blend with poor mechanical properties are, thus, generated due to the poor interfacial adhesion between the polymers. The interfacial adhesion between the thermoplastic and liquid crystalline polymers can be improved by the using functionalized CNTs. This helps in improving the mechanical properties of the polymer blend, along with developing superior electrical and chemical properties [120].

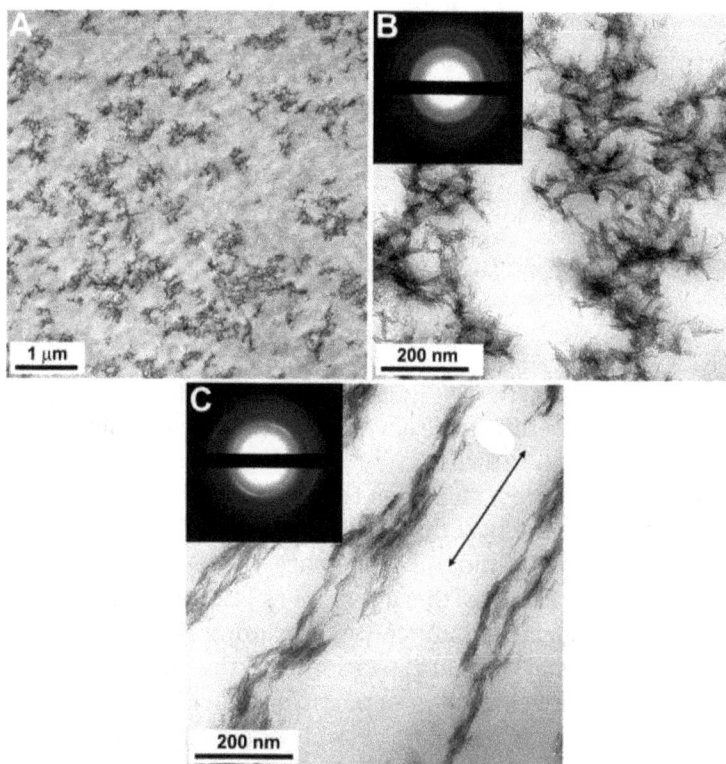

Figure 8.2 Transmission electron images of the composite containing 6.5 vol% of CdS nanorods before (A,B) and after the uniaxial deformation (C). Reproduced from Reference 94 with permission from American Chemical Society.

Mrozek and Taton [121] reported the LCP materials by seeding the growth of the LCP domains with the single-wall carbon nanotubes (SWNTs). The application of an electric field across the polyoxazoline melt incorporated with SWNTs seeds resulted in the orientation of the seeds (Figure 8.3). Subsequent cooling of the material resulted in the domains oriented in the direction of the applied field. Only a small amount of CNTs was required due to their action as nucleating agents.

Scheme 1

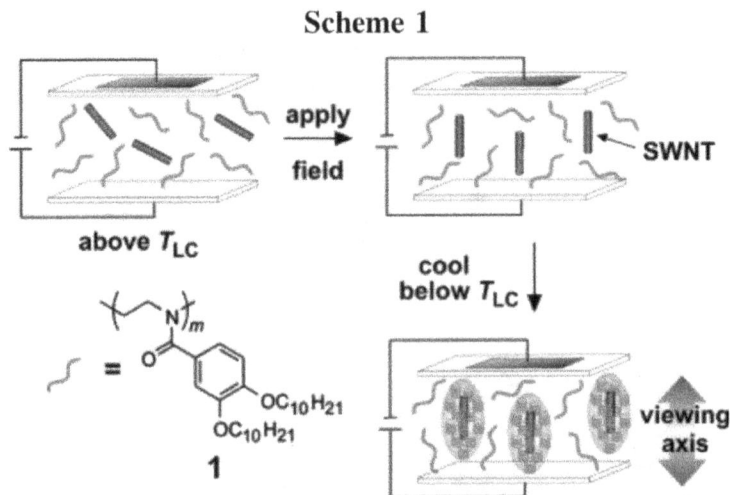

Figure 8.3 Schematic of the synthesis of the nanohybrid. Reproduced from Reference 121 with permission from American Chemical Society.

Display technologies have been the prime users of the liquid crystalline materials in the past decade. Bio-medical devices, light modulators, sensors and organic transistors are some of the other areas of commercial applications of these polymers [122-130]. In most of the LCP composites, surface modification of nanoparticles is required for generating interfacial compatibilization with the polymer phase. This paves the way of developing nanoparticles which are coated with LC ligands. The ligand exchange process is widely used for this purpose [130-136].

A number of studies have reported the *in-situ* synthesis of nanoparticles in the LC matrices for the purpose of developing nanocomposites. This involves the subsequent modifications of the precursors using techniques such as hydrolysis, oxidization and reduction. However, this approach faces challenges with respect to the compatibility of the precursor with the liquid crystal phase [136-151].

8.5 Conclusion

Many liquid crystalline polymer nanocomposite systems, exhibiting superior structural properties, have been developed in the recent past. Though many existing challenges currently hinder the full utilization of the potential of such materials for a variety of commercial applications, however, the ongoing research efforts in this direction are expected to further expand their widespread use in the near future.

References

1. *Liquid Crystalline Polymers: Synthesis, Properties and Applications*, Mittal, V. (ed.), Central West Publishing, Australia (2018).
2. Khoo, I.-C., and Wu, S.-T (1993) *Optics and Nonlinear Optics of Liquid Crystals*, World Scientific, Singapore.
3. *Recent Advances in Liquid Crystalline Polymers*, Chapoy, L. L. (ed.), Springer, Germany (1985).
4. Blumstein, A. (1985) *Polymeric Liquid Crystals*, Springer, Germany.
5. Griffin, A. C. (1984) *Liquid Crystals and Ordered Fluids*, 4th volume, Springer, Germany.
6. Xu, Q.-W., Man, H.-C. and Lau, W.-S. (1998) Study on the morphology and mechanical properties of binary and ternary blends of semiaromatic LCP/PP/PC. *Polymer-Plastics Technology and Engineering*, **37**, 253-259.
7. Lizuka, E. (2000) Effects of magnetic fields on polymer liquid crystals. *International Journal of Polymeric Materials*, **45**, 191-238.
8. Broer, D. J., Finkelmann, H., and Kondo, K. (1988) In-situ photopolymerization of an oriented liquid-crystalline acrylate. *Macromolecular Chemistry and Physics*, **189**, 185-194.
9. Broer, D. J., Mol, G. N., and Challa, G. (1989) In situ photopolymerization of an oriented liquid-crystalline acrylate, 2. *Macromolecular Chemistry and Physics*, **190**, 19-30.
10. Broer, D. J., Boven, J., Mol, G. N., and Challa, G. (1989) In-situ photopolymerization of oriented liquid-crystalline acrylates, 3. Oriented polymer networks from a mesogenic diacrylate. *Macromolecular Chemistry and Physics*, **190**, 2255-2268.
11. Hikmet, R. A. M., and Lub, J. (1995) Anisotropic networks with stable dipole orientation obtained by photopolymerization in the ferroelectric state. *Journal of Applied Physics*, **77**, 6234-6238.
12. He, L., Zhang, S., Jin, S., and Qi, Z. (1995) In-situ photopolymerization of an oriented chiral liquid crystal acrylate in an electric field. *Polymer Bulletin*, **34**, 7-12.
13. Hoyle, C. E., Mathias, L. J., Jariwala, C., and Sheng, D. (1996) Photopolymerization of a semifluorinated difunctional liquid crystalline monomer in a smectic phase. *Macromolecules*, **29**, 3182-3187.

14. Guymon, C. A., Hoggan, E. N., Clark, N. A., Rieker, T. P., Walba, D. M., and Bowman, C. N. (1997) Effects of monomer structure on their organization and polymerization in a smectic liquid crystal. *Science*, **275**, 57-59.

15. Broer, D. J., Mol, G. N., and Challa, G. (1991) Temperature effects on the kinetics of photoinitiated polymerization of dimethacrylates. *Polymer*, **32**, 690-695.

16. Hoyle, C. E., Kang, D., Jariwala, C., and Griffin, A. C. (1993) Efficient polymerization of a semi-fluorinated liquid crystalline methacrylate. *Polymer*, **34**, 3070-3075.

17. Hoyle, C. E., and Watanabe, T. (1994) Kinetics of polymerization of liquid-crystalline monomers: an exotherm and light scattering analysis. *Macromolecules*, **27**, 3790-3796.

18. Hoyle, C. E., Watanabe, T., and Whitehead, J. B. (1994) Anisotropic network formation by photopolymerization of liquid crystal monomers in a low magnetic field. *Macromolecules*, **27**, 6581-6588.

19. Sahlen, F., Trollsas, M., Hult, A., and Gedde, U. W. (1996) Synthesis and characterization of bifunctional liquid-crystalline monomers showing smectic C phase. photopolymerization and crosslinking. *Chemistry of Materials*, **8**(2), 382-388.

20. Hellermark, C., Gedde, U. W., and Hult, A. (1992) Synthesis and characterization of poly(11-(4'-cyano-trans-4-stilbenyloxy)undecanyl vinyl ether). *Polymer Bulletin*, **28**, 267-274.

21. Andersson, H., Gedde, U. W., and Hult, A. (1992) Preparation of ordered, crosslinked and thermally stable liquid crystalline poly(vinyl ether) films. *Polymer*, **33**, 4014-4018.

22. Broer, D. J., Lub, J., and Mol, G. N. (1993) Synthesis and photopolymerization of a liquid-crystalline diepoxide. *Macromolecules*, **26**, 1244-1247.

23. Jahromi, S., Lub, J., and Mol, G. N. (1994) Synthesis and photoinitiated polymerization of liquid crystalline diepoxides. *Polymer*, **35**, 622-629.

24. Andersson, H., Trollsas, M., Gedde, U. W., and Hult, A. (1995) Preparation of ordered and crosslinked films from liquid crystalline p-vinylphenoxy-based monomers. *Macromolecular Chemistry and Physics*, **196**, 3667-3676.

25. D'allest, J. F., Sixou, P., Blumstein, A., and Blumstein, R. B. (1988) Investigation of the nematic-isotropic biphase in thermotropic main chain polymers. Homogeneity of the pure nematic and isotropic phases. Part I: Microscopy and fractionation. *Molecular Crystals and Liquid Crystals Incorporating Nonlinear Optics*, **157**, 229-251.

26. Collyer A. A. (1996) Introduction to liquid crystal polymers. In: *Rheology and Processing of Liquid Crystal Polymers*, Acierno, D., Collyer, A. A. (eds), 2nd volume, Springer, Netherlands, doi: 10.1007/978-94-009-1511-4_1.

27. Thomas, E. L., Wood, B. A. (1985) Mesophase texture and defects in thermotropic liquid-crystalline polymers. *Faraday Discussions of the Chemical Society*, **79**, 229-239.

28. Vyas, R., Rida, A., Bhattacharya, S., and Tentzeris, M. M. (2007) Liquid Crystal Polymer (LCP): The Ultimate Solution for Low-Cost RF Flexible Electronics and Antennas. *Proceedings of the Polymers in Defense and Aerospace Applications*, France, p. 21.

29. Chung, T. S., Calundann, G. W., and East, A. J. (1989) Liquid-crystalline polymers and their applications. In: *Encyclopedia of Engineering Materials*, Chermisinoff, N. P. (ed.), Marcel Dekker, USA.

30. *High Performance Fibers*, Hearle, J. W. S. (ed.), Woodhead Publishing, USA (2001).

31. Carracher, Jr., C. E. (2017) *Polymer Chemistry*, 10th edition, CRC Press, USA.

32. *Liquid Crystals (Topics in Physical Chemistry)*, Stegemeyer, H. (ed.), Springer, Germany (1994).

33. *Liquid Crystals and Plastic Crystals*, Gray, G. W., and Winsor, P. A. (eds.), Ellis Horwood, UK (1974).

34. Robinson, C. (1956) Liquid-crystalline structures in solutions of a polypeptide. *Transactions of the Faraday Society*, **52**, 571-592.

35. Hijo, A. A. C. T., Maximo, G. JCosta, M. C., Cunha, R. L., Pereira, J. F. B., Kurnia, K. A., Batista, E. A. C., and Meirelles, A. J. A. (2017) Phase behavior and physical properties of new biobased ionic liquid crystals. *The Journal of Physical Chemistry B*, **121**(14), 3177-3189.

36. Ball, Z. T., Sivula, K., and Frechet, J. M. J. (2006) Well-defined fullerene-containing homopolymers and diblock copolymers with high fullerene content and their use for solution-phase and bulk organization. *Macromolecules*, **39**, 70-72.

37. Berg, S., Krone, V., and Ringsdorf, H. (1986) Structural variations of liquid-crystalline polymers: crossshaped and laterally linked mesogens in main chain and side group polymers. *Macromolecular Rapid Communications*, **7**, 381-388.

38. Chai, C. P., Zhu, X. Q., Wang, P., Ren, M. Q., Chen, X. F., Xu, Y. D., Fan, X. H., Ye, C., Chen, E. Q., and Zhou, Q. F. (2007) Synthesis and phase structures of mesogen-jacketed liquid crystalline polymers containing 1,3,4-oxadiazole based side chains. *Macromolecules*, **40**, 9361-9370.

39. Chandrasekhar, S., Sadashiva, B. K., and Suresh, K. A. (1977) Liquid crystals of disc-like molecules. *Pramana*, **9**(5), 471-480.

40. Chen, X., Tenneti, K. K., Li, C. Y., Bai, Y., Wan, X., Fan, X., Zhou, Q.-F., Rong, L., and Hsiao, B. S. (2007) Sidechain liquid crystalline poly(meth)acrylates with bent-Core mesogens. *Macromolecules*, **40**, 840-848.

41. Chen, X. F., Shen, Z., Wan, X. H., Fan, X. H., Chen, E. Q., Ma, Y., and Zhou, Q. F. (2010) Mesogen-jacketed liquid crystalline polymers.

Chemical Society Reviews, **39**, 3072-3101.

42. Godovsky, Y. K., Papkov, V. S., and Dusek, K. (1989) Thermotropic mesophases in element-organic polymers. *Advances in Polymer Science*, **88**, doi: 10.1007/BFb0017966.

43. Gopalan, P., Andruzzi, L., Li, X. F., and Ober, C. K. (2002) Fluorinated mesogen-jacketed liquid-crystalline polymers as surface-modifying agents: Design, synthesis and characterization. *Macromolecular Chemistry and Physics*, **203**, 1573-1583.

44. Mark. H. F. (1988) *Encyclopedia of Polymer Science and Engineering*, 9th volume, Wiley, USA.

45. Hessel, F., and Finkelmann, H. (1985) A new class of liquid crystal side chain polymers mesogenic groups laterally attached to the polymer backbone. *Polymer Bulletin*, **14**, 375-378.

46. Keith, C., Reddy, R. A., and Tschierske, C. (2005) The first example of a liquid crystalline side-chain polymer with bent-core mesogenic units: ferroelectric switching and spontaneous achiral symmetry breaking in an achiral polymer. *Chemical Communications*, **7**, 871-873.

47. Keith, C., Dantlgraber, G., Amaranatha Reddy, R., Baumeister, U., and Tschierske, C. (2007) Ferroelectric and antiferroelectric smectic and columnar liquid crystalline phases formed by silylated and non-silylated molecules with fluorinated bent cores. *Chemistry of Materials*, **19**(4), 694-710.

48. Chen, W.-H., Chuang, W.-T., Jeng, U.-S., Sheu, H.-S., and Lin, H.-C. (2011) New SmCG phases in a hydrogen-bonded bent-core liquid crystal featuring a branched siloxane terminal group. *Journal of American Chemical Society*, **133**(39), 15674-15685.

49. Westphal, E., Gallardo, H., Sebastian, N., Eremin, A., Prehm, M., Alaasar, M., and Tschierske, C. (2019) Liquid crystalline self-assembly of 2,5-diphenyl-1,3,4-oxadiazole based bent-core molecules and the influence of carbosilane end-groups. *Journal of Materials Chemistry C*, in print.

50. Jakli, A., Lavrentovich, O. D., and Selinger, J. V. (2018) Physics of liquid crystals of bent-shaped molecules. *Reviews of Modern Physics*, **90**, 045004.

51. Katranchev, B., and Petrov, M. (2016) Phase transitions in nanocomposites of hydrogen-bonded dimeric liquid crystals with mesogenic and non-mesogenic components. *Phase Transitions*, **89**(2), 115-132.

52. Pugh, C., Bae, J. Y., Dharia, J., Ge, J. J., and Cheng, S. Z. D. (1998) Induction of smectic layering in nematic liquid crystals using immiscible components. 2. laterally attached side-chain liquid-crystallin poly(norbornene)s and their low-molar-mass analogues with hydrocarbon/oligodimethylsiloxane substituents. *Macromolecules*, **31**, 5188-5200.

53. Saravanan, C., and Kannan, P. (2010) Fluorine-substituted azoben-

zene destabilizes polar form of optically switchable fulgimide unit in copolymer system. *Journal of Polymer Science, Part A: Polymer Chemistry*, **48**, 1565-1578.

54. Saravanan, C., Senthil, S., and Kannan, P. (2008) Click chemistry-assisted triazole-substituted azobenzene and fulgimide units in the pendant-based copoly(decyloxymethacrylate)s for dual-mode optical switches. *Journal of Polymer Science, Part A: Polymer Chemistry*, **46**, 7843-7860.

55. Vix, A., Stocker, W., Stamm, M., Wilbert, G., Zentel, R., and Rabe, J. P. (1998) Chain folding in liquid-crystalline main-chain polymers with a smectic phase. *Macromolecules*, **31** (26), 9154-9156.

56. Reck, B., and Rangsdorf, H. (1985) Combined liquid crystalline polymers, mesogens in the main chain and as side groups. *Macromolecular Rapid Communications*, **6**, 291-299.

57. Reddy, R. A., and Sadashiva, B. K. (2003) Influence of fluorine substituent on the mesomorphic properties of five-ring ester banana-shaped molecules. *Liquid Crystals*, **30**, 1031-1050.

58. Reddy, G. S. M., Narasimhaswamy, T., Jayaramudu, J., Sadiku, E. R., Raju, K. M., Ray, S. S. (2013) A new series of two-ring-based side chain liquid crystalline polymers, synthesis and mesophase characterization. *Australian Journal of Chemistry*, **66**, 667-675.

59. Tenneti, K. K., Chen, X., Li, C. Y., Shen, Z., Wan, F. X., Zhou, Q. F., Rong, L., and Hsiao, B. S. (2009) Influence of LC content on the phase structures of side-chain liquid crystalline block copolymers with bent-core mesogens. *Macromolecules*, **42**, 3510-3517.

60. Komori, T., and Shinkai, S. (1993) Novel columnar liquid crystals designed from cone-shaped calix[4] arenes. The rigid bowl is essential for the formation of the liquid crystal phase. *Chemistry Letters*, **22**, 1455-1458.

61. Kostromin, S. G., Shibaev, V. P., and Plate, N. A. (1987) Thermotropic liquid-crystalline polymers XXVI. Synthesis of comb-like polymers with oxygen containing spacers and a study of their phase transitions. *Liquid Crystals*, **2**, 195-200.

62. Lecommandoux, S., Klok, H. A., Sayar, M., and Stupp, S. I. (2003) Synthesis and self-organization of rod–dendron and dendron–rod–dendron molecules. *Journal of Polymer Science, Part A: Polymer Chemistry*, **41**, 3501-3518.

63. Lehn, J. M. (1988) Supramolecular chemistry, scope and perspectives molecules supramolecular and molecular devices. *Angewandte Chemie, International Edition*, **27**, 89-112.

64. Tamm, L. K., and McConnell, H. M. (1985) Supported phospholipid bilayers. *Biophysical Journal*, **47**(1), 105-113.

65. Maier, G. (2001) Low dielectric constant polymers for microelectronics. *Progress in Polymer Science*, **26**, 3-65.

66. Matsuo, Y., Muramatsu, A., Hamasaki, R., Mizoshita, N., Kato, T., and

Nakamura, E. (2004) Regioselective synthesis of 1,4-Di(organo)[60]fullerenes through DMF-assisted monoaddition of silylmethyl grignard reagents and subsequent alkylation reaction. *Journal of American Chemical Society*, **126**, 432-433.

67. Niori, T., Sekine, T., Watanabe, J., Furukawa, T., and Takezoe, H. (1996) Distinct ferroelectric smectic liquid crystals consisting of banana shaped achiral molecules. *Journal of Materials Chemistry*, **6**, 1231-1233.

68. Noel, C., and Navarad, P. (1991) Liquid crystal polymers. *Progress in polymer science*, **16**, 55-110.

69. Ortega, J., Folcia, C. L., Etxebarria, J., Gimeno, N., and Ros, M. B. (2003) Interpretation of unusual textures in the B2 phase of a liquid crystal composed of bent-core molecules. *Physical Review E*, **68**, 011707.

70. Shibaev, V. P. (2009) Liquid–crystalline polymers, past, present, and future. *Polymer Science Series A*, **51**, 1131.

71. Shibaev, V. P., and Plate, N. A. (1984) Thermotropic liquid-crystalline polymers with mesogenic side groups. *Advances in Polymer Science*, **60/61**, doi: 10.1007/3-540-12994-4_4.

72. Shibaev, V. P., and Plate, N. A. (1985) Synthesis and structure of liquid-crystalline side-chain polymers. *Pure and Applied Chemistry*, **57**, 1589-1602.

73. Spassky, N., Lacoudre, N., Le Borgne, A., Varion, J. P., Jun, C. L., Friedrich, C., and Noel, C. (1989) Liquid crystal polymers with terminally 1- phenyl 2-(4-cyanophenyl)-ethane substituted side chains. *Macromolecular Symposia*, **24**, 271-281.

74. Sawamura, M., Kawai, K., Matsuo, Y., Kanie, K., Kato, T., and Nakamura, E. (2002) Stacking of conical molecules with a fullerene apex into polar columns in crystals and liquid crystals. *Nature*, **419**, 702-705.

75. Siva Mohan Reddy G., Jayaramudu J., Ray S. S., Varaprasad K., and Rotimi Sadiku E. (2015) Side chain liquid crystalline polymers: Advances and applications. In: *Liquid Crystalline Polymers*, Thakur, V., and Kessler, M. (eds.), Springer, Germany, doi: 10.1007/978-3-319-20270-9_16.

76. Ahn, S.-k., Deshmukh, P., Gopinadhan, M., Osuji, C. O., and Kasi, R. M. (2011) Side-chain liquid crystalline polymer networks, exploiting nanoscale smectic polymorphism to design shape-memory polymers. *ACS Nano*, **5**(4), 3085-3095.

77. Percec, V., and Yourd, R. (1998) Liquid crystalline polyethers and copolyethers based on conformational isomerism. 3. The influence of thermal history on the phase transitions of the thermotropic polyethers and copolyethers based on 1-(4-hydroxyphenyl)-2-(2-methyl-4-hydroxyphenyl)ethane and flexible spacers containing an odd number of methylene units. *Macromolecules*, **22**, 3229–3242.

78. Percec, V., Rodrguez-Prarada, J. M., and Ericsson, C. (1987) Synthesis and characterization of liquid crystalline poly (p-vinylbenzyl ether)s. *Polymer Bulletin*, **17**, 347-352.

79. Percec, V., Heck, J., and Ungar, G. (1991) Liquid-crystalline polymers containing mesogenic units based on half-disk and rodlike moieties. 5. side-chain liquid-crystalline poly(methylsiloxanes) containing hemiphasmidic mesogens based on 4-[[3,4,5-tris(alkan-l-yloxy) benzoyl]oxy]-4'- [[p-(propan-1-yloxy)-benzoyl] oxy] biphenyl groups. *Macromolecules*, **24**, 4957-4962.

80. Percec, V., Heck, J., Lee, M., Ungar, G., and Alvarez-Castillo, A. (1992) Poly{2-vinyloxyethyl 3,4,5-tris [4-(n-dodecanyloxy)ben-zyloxy]benzoate}, a self-assembled supramolecular polymer similar to tobacco mosaic virus. *Journal of Materials Chemistry*, **2**, 1033-1039.

81. Percec, V., Schlueter, D., Ronda, J. C., Johansson, G., Ungar, G., and Zhou, J. P. (1996) Tubular architectures from polymers with tapered side groups. assembly of side groups via a rigid helical chain conformation and flexible helical chain conformation induced via assembly of side groups. *Macromolecules*, **29**, 1464-1472.

82. Percec, V., Ahn, C. H., Ungar, G., Yeardley, D. J. P., Moller, M., and Sheiko, S. S. (1998) Controlling polymer shape through the self-assembly of dendritic side-groups. *Nature*, **391**, 161-164.

83. Percec, V., Holerca, M. N., Uchida, S., Cho, W. D., Ungar, G., Lee, Y., and Yeardley, D. J. P. (2002) Exploring and expanding the three-dimensional structural diversity of supramolecular dendrimers with the aid of libraries of alkali metals of their AB3 minidendritic carboxylates. *Chemistry, A European Journal*, **8**, 1106-1117.

84. Percec, V., Rudick, J. G., Peterca, M., Wagner, M., Obata, M., Mitchell, C. M., Cho, W. D., Balagurusamy, V. S. K., and Heiney, P. A. (2005) Thermoreversible cis-cisoidal to cis-transoidal isomerization ofhelical dendronized polyphenylacetylenes. *Journal of American Chemical Society*, **127**, 15257-15264.

85. Piunova, V. A., Miyake, G. M., Daeffler, C. S., Weitekamp, R. A., and Grubbs, R. H. (2013) Highly ordered dielectric mirrors via the self-assembly of dendronized block copolymers. *Journal of American Chemical Society*, **135**, 15609-15616.

86. Pugh, C., and Schrock. R. R. (1992) Synthesis of side-chain liquid crystal polymers by living ring-opening metathesis polymerization. 3. influence of molecular weight, interconnecting unit, and substituent on the mesomorphic behavior of polymers with laterally attached mesogens. *Macromolecules*, **25**, 6593-6604.

87. Zorn, M., Meuer, S., Tahir, M. N., Khalavka, Y., Sönnichsen, C., Tremel, W., and Zentel, R. (2008) Liquid crystalline phases from polymer functionalised semiconducting nanorods. *Journal of Materials Chemistry*, **18**, 3050-3058.

88. Riou, O., Lonetti, B., Davidson, P., Tan, R. P., Cormary, B., Mingotaud, A.-F., Di Cola, E., Respaud, M., Chaudret, B., Soulantica, K., and Mauzac, M. (2014) Liquid crystalline polymer–Co nanorod hybrids, structural analysis and response to a magnetic field. *The Journal of Physical Chemistry B*, **118** (11), 3218-3225.

89. Zorn, M., and Zentel, R. (2008) Liquid crystalline orientation of semiconducting nanorods in a semiconducting matrix. *Macromolecular Rapid Communications*, **29**, 922-927.

90. Meuer, S., Oberle, P., Theato, P., Tremel, W. and Zentel, R. (2007) Liquid crystalline phases from polymer-functionalized TiO$_2$ nanorods. *Advanced Materials*, **19**, 2073-2078.

91. Li, L.-S., and Alivisatos, A. P. (2003) Semiconductor nanorod liquid crystals and their assembly on a substrate. *Advanced Materials*, **15**, 408-411.

92. Yoshino, T., Kondo, M., Mamiya, J.-I., Kinoshita, M., Yu, Y. and Ikeda, T. (2010) Three-dimensional photomobility of crosslinked azobenzene liquid-crystalline polymer fibers. *Advanced Materials*, **22**, 1361-1363.

93. Jang, J., and Bae, J. (2005) formation of polyaniline nanorod/liquid crystalline epoxy composite nanowires using a temperature-gradient method. *Advanced Function Materials*, **15**, 1877-1882.

94. Ezhov, A. A., Shandryuk, G. A., Bondarenko, G. N., Merekalov, A. S. Abramchuk, S. S., Shatalova, A. M., Manna, P., Zubarev, E. R., and Talroze, R. V. (2011) liquid-crystalline polymer composites with CdS nanorods, structure and optical properties. *Langmuir*, **27**(21), 13353-13360.

95. Nikhil, R., Gearheart, J. L. A., Obare, S. O., Johnson, C. J., Edler, K. J., Mann S., and Murphy, C. J. (2002) Liquid crystalline assemblies of ordered gold nanorods. *Journal of Materials Chemistry*, **12**, 2909-2912.

96. Dessombz, A., Chiche, D., Davidson, P., Panine, P., Chanéac, C., and, and Jolivet, J.-P. (2007) design of liquid-crystalline aqueous suspensions of rutile nanorods, evidence of anisotropic photocatalytic properties. *Journal of American Chemical Society*, **129**(18), 5904-5909.

97. Horsch, M. A., Zhang, Z., and Glotzer, S. C. (2005) Self-assembly of polymer-tethered nanorods. *Physical Review Letters*, **95**(5), 056105.

98. Meuer, S. Fischer, K. Mey, I. Janshoff, A. Schmidt, M. and Zentel, R. (2008) Liquid crystals from polymer-functionalized TiO$_2$ nanorod mesogens. *Macromolecules*, **41**(21), 7946-7952.

99. Guo, L., Cheng, J. X., Li, X.-Y., Yan, Y. J., Yang, S. H., Yang, C. L., Wang, J. N., and Ge, W. K. (2001) Synthesis and optical properties of crystalline polymer-capped ZnO nanorods, *Materials Science and Engineering C*, **16**(1-2), 123-127.

100. *Liquid Crystal Polymers*, Plate, N. A. (ed.), Plenum Press, USA (1992).
101. *Advances in Liquid Crystals*, Brown, G. H. (ed.), Academic Press, USA (1975).
102. Collings, P. J. (1990) *Liquid Crystals, Nature's Delicate Phase of Matter*, Princeton University Press, USA.
103. Tjong, S. C. (2003) Structure, morphology, mechanical and thermal characteristics of the in situ composites based on liquid crystalline polymers and thermoplastics. *Materials Science and Engineering R: Reports*, **41**, 1-60.
104. Donald, A. M., and Windle, A. H. (1992) *Liquid Crystalline Polymers*, Cambridge University Press, UK.
105. *Surface Modification of Nanotube Fillers*, Mittal, V. (ed.), Wiley VCH, Germany (2011).
106. *Polymer-Graphene Nanocomposites*, Mittal, V. (ed.), Royal Society of Chemistry, UK, 2012.
107. Iijima, S. (1991) Helical microtubules of graphitic carbon. *Nature*, **354**, 56-58.
108. Demus, D., Gray, G. W., Speiss, H. W., Goodby, J. W., and Vill, V. (1998) *Handbook of Liquid Crystals*, Wiley-VCH, Germany.
109. Park, S. K., Kim, S. H., and Hwang, J. T. (2008) Carboxylated multiwall carbon nanotube-reinforced thermotropic liquid crystalline polymer nanocomposites. *Journal of Applied Polymer Science*, **109**, 388-396.
110. Kim, J. Y., Kim, D. K., and Kim, S. H. (2009) Effect of modified carbon nanotube on physical properties of thermotropic liquid crystal polyester composites. *European Polymer Journal*, **45**, 316-324.
111. Kaito, A., Kyotani, M., and Nakayama, K. (1990) Effects of annealing on the structure formation in a thermotropic liquid crystalline copolyester. *Macromolecules*, **23**, 1035-1040.
112. Moniruzzaman, M., and Winey, K. I. (2006) Polymer nanocomposites containing carbon nanotubes. *Macromolecules*, **39**, 5194-5205.
113. Xie, X. L., Mai, Y. W., and Zhou, X. P. (2005) Dispersion and alignment of carbon nanotubes in polymer matrix: A review. *Materials science and Engineering R: Reports*, **49**, 89-112.
114. Coleman, J. N., Khan, U., Blau, W. J., and Gun'ko, Y. K. (2006) Small but strong, A review of the mechanical properties of carbon nanotube-polymer composites. *Carbon*, **44**, 1624-1652.
115. Coleman, J. N., Khan, U., and Gun'ko, Y. K. (2006) Mechanical reinforcement of polymers using carbon nanotubes. *Advanced Materials*, **18**, 689-706.
116. Kiss, G. (1987) In situ composites, Blends of isotropic polymers and thermotropic liquid crystalline polymers. *Polymer Engineering and Science*, **27**, 410-423.
117. Lin, Y. G., and Winter, H. H. (1991) High-temperature recrystalliza-

tion and rheology of a thermotropic liquid crystalline polymer. *Macromolecules*, **24**, 2877-2882.

118. Serpe, G., and Economy, J. (1992) Ordering processes in the 2,6-hydroxynaphthoic acid (HNA) rich copolyesters of p-hydroxybenzoic acid and HNA. *Macromolecular Symposia*, **53**, 65-75.

119. Cheng, H. K. F., Sahoo, N. G., Li, L., Chan, S. H., and Zhao, J. (2010) Molecular interactions in PA6, LCP and their blend incorporated with functionalized carbon nanotubes. *Key Engineering Materials*, **447-448**, 634-638.

120. Cheng, H. K. F., Basu, T., Sahoo, N. G., Li, L., and Chan, S. H. (2012) Current advances in the carbon nanotube/thermotropic main-chain liquid crystalline polymer nanocomposites and their blends. *Polymers*, **4**(2), 889-912.

121. Mrozek, R. A., and Taton, T. A. (2005) Alignment of liquid-crystalline polymers by field-oriented, carbon nanotube directors. *Chemistry of Materials*, **17**(13), 3384-3388.

122. Dasgupta, D., Shishmanova, I. K., Ruiz-Carretero, A., Lu, K., Verhoeven, M., van Kuringen, H.P.C, Portale, G., Leclere, P., Bastiaan-sen, C. W. M., Broer, D. J., and Schenning, A. P. H. J. (2013) Patterned silver nanoparticles embedded in a nanoporous smectic liquid crystalline polymer network. *Journal of American Chemical Society*, **135**, 10922-10925.

123. Dobbs, W., Suisse, J. M., Douce, L., and Welter, R. (2006) Electrodeposition of silver particles and gold nanoparticles from ionic liquid-crystal precursors. *Angewandte Chemie International Edition*, **45**, 4179-4182.

124. Domenici, V., Zupancic, B., Laguta, V. V., Belous, A. G., V'Yunov, O. I., Remskar, M., Zalar, B. (2010) PbTiO3 nanoparticles embedded in a liquid crystalline elastomer matrix, structural and ordering properties. *Journal of Physical Chemistry C*, **114**, 10782-10789.

125. Domenici, V., Conradi, M., Remskar, M., Virsek, M., Zupancic, B., Mrzel, A., Chambers, M., and Zalar, B. (2011) New composite films based on MoO3-x nanowires aligned in a liquid single crystal elastomer matrix. *Journal of Materials Science*, **46**, 3639-3645.

126. Saliba, S., Coppel, Y., Achard, M. F., Mingotaud, C., Marty, J. D., and Kahn, M. L. (2011) Thermotropic liquid crystals as templates for anisotropic growth of nanoparticles. *Angewandte Chemie, International Edition*, **50**, 12032-12035.

127. Saliba, S., Coppel, Y., Mingotaud, C., Marty, J. D., and Kahn, M. L. (2012) ZnO/liquid crystalline nanohybrids, from solution properties to the control of the anisotropic growth of nanoparticles. *Chemistry, A European Journal*, **18**, 8084-8091.

128. Saliba, S., Mingotaud, C., Kahn, M. L., and Marty, J. D. (2013) Liquid crystalline thermotropic and lyotropic nanohybrids. *Nanoscale*, **5**, 6641-6661.

129. Shandryuk, G. A., Matukhina, E. V., Vasil'ev, R. B., Rebrov, A., Bondarenko, G. N., Merekalov, A. S., Gas'kov, A. M., and Talroze, R. V. (2008) Effect of H-bonded liquid crystal polymers on CdSe quantum dot alignment within nanocomposite. *Macromolecules*, **41**, 2178-2185.

130. Taubert, A. (2004) CuCl nanoplatelets from an ionic liquid-crystal precursor. *Angewandte Chemie, International Edition*, **43**, 5380-5382.

131. Vasilets, V. N., Savenkov, G. N., Merekalov, A. S., Shandryuk, G. A., Shatalova, A. M., and Tal'roze, R. V. (2011) Immobilization of quantum dots of cadmium selenide on the matrix of a graft liquid-crystalline polymer. *Polymer Science Series A*, **53**, 521-526.

132. Zadoina, L., Lonetti, B., Soulantica, K., Mingotaud, A. F., Respaud, M., Chaudret, B., and Mauzac, M. (2009) Liquid crystalline magnetic materials. *Journal of Materials Chemistry*, **19**, 8075-8078.

133. Zadoina, L., Soulantica, K., Ferrere, S., Lonetti, B., Respau, M., Mingotaud, A. F., Falqui, A., Genovese, A., Chaudret, B., and Mauzac, M. (2011) In situ synthesis of cobalt nanoparticles in functionalized liquid crystalline polymers. *Journal of Materials Chemistry*, **21**, 6988-6994.

134. Barmatov, E. B., Pebalk, D. A., and Barmatova, M. V. (2004) Influence of silver nanoparticles on the phase behavior of side-chain liquid crystalline polymers. *Langmuir*, **20**, 10868-10871.

135. Brochard, F., and de Gennes, P. G. (1970) Theory of magnetic suspensions in liquid crystals. *Journal de Physique*, **31**(7), 691-708.

136. Chambers, M., Zalar, B., Remskar, M., Zumer, S., and Finkelmann, H. (2006) Actuation of liquid crystal elastomers reprocessed with carbon nanoparticles. *Applied Physics Letters*, **89**, 243116.

137. Domracheva, N. E., Pyataev, A. V., Manapov, R. A., and Gruzdev, M. S. (2011) Magnetic resonance and M€ossbauer studies of superparamagnetic γ-Fe$_2$O$_3$ nanoparticles encapsulated into liquidcrystalline poly(propylene imine) dendrimers. *ChemPhysChem*, **12**, 3009-3019.

138. Garcia-Marquez, A., Demortiere, A., Heinrich, B., Guillon, D., Begin-Colin, S., and Donnio, B. (2011) Iron oxide nanoparticle-containing main-chain liquid crystalline elastomer, towards soft magnetoactive networks. *Journal of Materials Chemistry*, **21**, 8994-8996.

139. Gascon, I., Marty, J. D., Gharsa, T., and Mingotaud, C, (2005) Formation of gold nanoparticles in a sidechain liquid crystalline network, influence of the structure and macroscopic order of the material. *Chemistry of Materials*, **17**, 5228-5230.

140. Haberl, J. M., Sanchez-Ferrer, A., Mihut, A. M., Dietsch, H., Hirt, A. M., and Mezzenga, R. (2013) Liquid crystalline elastomer-nanoparticle hybrids with reversible switch of magnetic memory. *Advanced Materials*, **25**, 1787-1791.

141. Haberl, J. M., Sanchez-Ferrer, A., Mihut, A. M., Dietsch, H., Hirt, A. M., and Mezzenga, R. (2013) Straininduced macroscopic magnetic anisotropy from smectic liquid-crystalline elastomermaghemite nanoparticle hybrid nanocomposites. *Nanoscale*, **5**, 5539-5548.
142. Hegman, T., Qi, H., and Marx, V. M. (2007) Nanoparticles and liquid crystals, synthesis, self-assembly, defect formation and potential applications. *Journal of Inorganic and Organometallic Polymers and Materials*, **17**, 483-508.
143. Kaiser, A., Winkler, M., Krause, S., Finkelmann, H., and Schmidt, A. M. (2009) Magnetoactive liquid crystal elastomer nanocomposites. *Journal of Materials Chemistry*, **19**, 538-543.
144. Lee, J., and W., Jin, J. (2003) Formation of gold nanoparticles within a liquid crystalline polymeric matrix. *Journal of Nanoscience and Nanotechnology*, **3**, 219-221.
145. Li, F, Chen, W, and Chen, Y. W. (2012) Mesogen induced self-assembly for hybrid bulk heterojunction solar cells based on a liquid crystal D-A copolymer and ZnO nanocrystals. *Journal of Material Chemistry*, **22**, 6259-6266.
146. Mallia, V. A., Vemula, P. K., John, G., Kumar, A., and Ajayan, P. M. (2007) In situ synthesis and assembly of gold nanoparticles embedded in glass-forming liquid crystals. *Angewandte Chemie, International Edition*, **46**, 3269-3274.
147. Marshall, J. E., Ji, Y., Torras, N., Zinoviev, K., and Terentjev, E. M. (2012) Carbon-nanotube sensitized nematic elastomer composites for IR-visible photo-actuation. *Soft Matter*, **8**, 1570-1574.
148. Mertelj, A., Lisjak, D., Drofenik, M., and Copic, M. (2013) Ferromagnetism in suspensions of magnetic platelets in liquid crystal. *Nature*, **504**, 237-241.
149. Montazami, R., Spillmann, C. M., Naciri, J., and Ratna, B. R. (2012) Enhanced thermomechanical properties of a nematic liquid crystal elastomer doped with gold nanoparticles. *Sensors and Actuators A: Physical*, **178**, 175-178.
150. Nguyen, H. H., Valverde, S. C., Lavedan, P., Goudoune`che, D., Mingotaud, A. F., Lauth-de Viguerie, N., and Marty, J. D. (2014) Mesomorphic ionic hyperbranched polymers, effect of structural parameters on liquid-crystalline properties and on the formation of gold nanohybrids. *Nanoscale*, **6**, 3599-3610.
151. Riou, O., Zadoina, L., Lonetti, B., Soulantica, K., Mingotaud, A.-F., Respaud, M., Chaudret, B., and Mauzac, M. (2012) *In situ* and *ex situ* syntheses of magnetic liquid crystalline materials: A comparison. *Polymers*, **4**(1), 448-462.

Nanocomposites of Polypropylene, Polypropylene/
PP-g-MA and Polypropylene/Polyamide Blend with
Graphene and Boron Nitride: Thermal, Mechanical,
Morphological and Electrical Studies

9.1 Introduction

In the last two decades, polymer nanocomposites have evolved into the materials of high commercial relevance due to the significant enhancement in the polymer properties by the incorporation of small fractions of functional nanofillers [1]. Due to the presence of filler particles at the nanoscale, a large number of polymer-filler interfacial contacts develop in the materials, thus, resulting in effective transfer of load as well as heat.

Polypropylene (PP) is a stereo-regular polymer, with properties depending on the crystalline structure and various tacticities. Moreover, PP possesses various beneficial properties like low density, high thermal resistance, resistance to chemical attack (insoluble in almost all solvents at room temperature), ease of processing and recyclability. Interestingly, PP manifests mechanical profile closely to the engineering thermoplastics. Owing to this extraordinary set of properties, PP is a commonly used thermoplastic for structural applications [2]. A large variety of reinforcing fillers such as silica, clay, nanotubes and graphene have been explored as fillers in PP. Graphene, which consists of one atomic thick sheets of covalently sp^2-bonded carbon atoms in a hexagonal arrangement, has received immense attention of researchers for the generation of polymer nanocomposites [3,4]. Its choice as a filler is due to its excellent electrical and mechanical properties, which are significantly better than other inorganic filler materials. For instance, a single defect-free graphene layer has Young's modulus of ~1.0 TPa, intrinsic strength ~42 N/m, thermal conductivity ~4840-5300 W/(m.K), electron mobility exceeding 25,000 cm^2/V.s, excellent gas impermeability and specific surface area of

Mona Al Hosani, M. R. Vengatesan and Vikas Mittal*, The Petroleum Institute (part of Khalifa University of Science and Technology), Abu Dhabi, UAE
*Current address: Bletchington, Wellington County, Australia
© 2019 Central West Publishing, Australia

~2630 m^2/g [3]. In a recent study, Pawar *et al.* [5] reported functional PP/EVA-MA (ethylene vinyl acetate-*graft*-maleic anhydride) nanocomposites with amine functionalized graphite oxide, GrO-NH$_2$. As shown in Figure 9.1, the filler stacks were only a few layers thick and

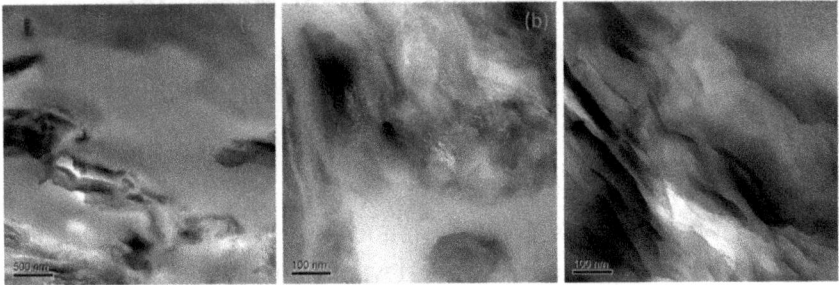

Figure 9.1 (a-c) Morphology of PP+EVA-MA+1%GrONH$_2$ composite. Reproduced from Reference 5.

were largely present in the EVA-MA phase. The storage modulus of the composites was also observed to be enhanced owing to the polymer-filler interactions and effective filler dispersion leading to higher stress transfer from the polymer chains to the filler platelets. In another study, Mittal *et al.* [6] reported blend nanocomposites of polypropylene and polystyrene with graphene and ethylene-co-acrylic acid (EAA) copolymer. The generated blend nanocomposites exhibited superior properties than the corresponding blends and are more useful alternatives than not only the parent materials but also other thermoplastics due to optimal properties. In addition, the morphology of the blend nanocomposites was also stable even after a few high temperature processing cycles, though the components were only kinetically mixed. Figure 9.2 demonstrates the effect of filler as well as compatibilizer on the tensile properties of the materials. Overall, even though the filler did not exfoliate extensively, the EAA addition still helped to achieve better composite performance by enhancing the overall extent of filler dispersion. In addition, the localization of filler in the PP phase also correspondingly affected the composite properties, as a function of blend composition.

Most of the polymers are insulators at macro-scale and usually have thermal conductivity below 0.5 W/(m.K). Generally, the polymers with higher degree of crystalline domains have higher thermal conductivity than the amorphous polymers. Similarly, the presence of side chains in the main chain also results in lower conductivity of the polymer. Specifically, in thermoplastic polymers, thermal conductivity is affected by the factors such as chain structure and orientation, crystallinity and interchain interactions. On the contrary, the factors which affect the thermal conductivity in thermoset polymers are mainly liquid crystal domain size and content, curing conditions

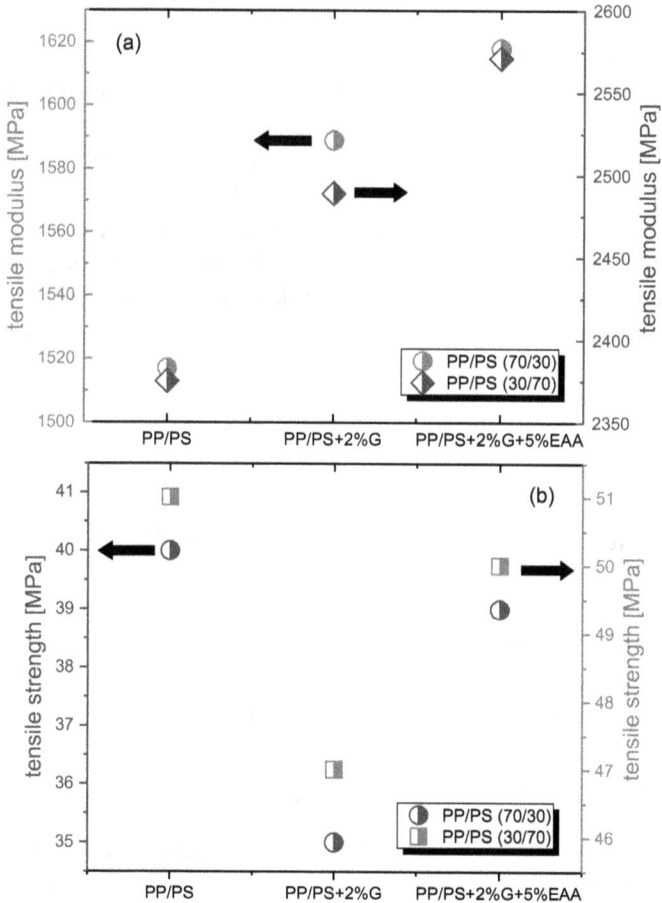

Figure 9.2 Tensile properties of (a) PP/PS (70/30) blend and its composites; (b) PP/PS (30/70) blend and its composites. Reproduced from Reference 6.

and orientation [7]. In addition to the mechanical and thermal properties, incorporation of fillers to the polymers is also beneficial to enhance the thermal and electrical conductivities. Usually, one dimensional fillers (1-D) develop well connected long conductive pathways in composite systems which result in significantly improved thermal conductivity. Examples of 1-D fillers are carbon nanotubes, carbon fibers, Si_3N_4 nanowires, boron nitride nanotubes, silver nanowires, copper nanowires, etc. [8-10]. Similarly, two dimensional (2-D) fillers, such as platelet fillers, are also considered as high aspect ratio fillers and can exhibit very high in-plane thermal conductivity, examples of such fillers with plate like morphology are boron nitride, graphene, Al_2O_3, TiB_2 and SiC [11].

In this work, PP nanocomposites have been developed with graphene as filler for achieving enhancement in mechanical and thermal properties, along with thermal and electrical conductivities. The use of polypropylene-*graft*-maleic anhydride (PP-g-MA) based compatibilizer as well as polyamide (PA) as blend component was employed for further enhancing the composite performance. In addition, boron nitride was also used as reinforcing filler so as to compare its performance with graphene for enhancing the polymer properties.

9.2 Experimental

9.2.1 Materials

Polypropylene (HD91SCF) was procured from Abu Dhabi Polymers Limited. Polyamide Durethan T40, PP-g-MA (M_w ~9,100, M_n ~3,900 and maleic anhydride % grafting 8-10%) and *o*-xylene were purchased from Sigma Aldrich. The materials were used as received, without further purification. Graphene (G) with a trade name N002-PDR was purchased from Angstron Materials, USA.

9.2.2 Preparation of Nanocomposites

The nanocomposites were prepared using solution blending technique. For instance, a 500 mL round bottom flask was added with 300 mL of xylene and 150 mg (3 wt%) of graphene. The suspension was kept for 1 h in ultra-sonication bath. 4.85 g of polypropylene was subsequently added to the flask. Afterwards, the flask was fitted with a reflux condenser in a silicon oil bath at 130 °C for 2 h. The solvent was subsequently evaporated at 100 °C in a vacuum atmosphere for around 12 h until the composite become fully dry. The prepared composite was then melted and mixed in the extruder for 15 min. The composites with PP-g-MA were generated by adding the PP-g-MA (fixed at 5 wt%) before the addition of PP. Similarly, the composites

with 5 wt% graphene were generated. By following the same procedure, the composites with 5 and 10 wt% BN were also generated. For
generating the blend composites with PA, a 50:50 PP/PA blend was
used. For comparison, PP/PP-g-MA and PP/PA were also prepared
using the same procedure. In addition, pure PP was also subjected to
same processing protocols so as to obtain accurate property compar-
isons among the materials. The disc and dumbbell-shaped test samples were prepared by mini injection molding machine (HAAKE Mini-
Jet, Germany) at a processing temperature of 190 °C. The injection
pressure was 1000 bar for 6 s, whereas holding pressure was 400 bar
for 3 s. The temperature of the mold was kept at 55 °C.

9.2.3 Characterization

Differential scanning calorimetric analysis of the materials was carried out using Discovery DSC from TA Instruments under nitrogen atmosphere. The scans were obtained from 30 to 200 °C (first heating),
200 to -50 °C (first cooling) and -50 to 200 °C (second hating) using
heating and cooling rates of 10 °C/min and 5 °C/min, respectively.
Second heating runs were used for the comparison purpose. Thermal
degradation properties of the polymer, blends and nanocomposites
were recorded using Discovery thermogravimetric analyzer (TGA)
from TA Instruments. Nitrogen was used as a carrier gas and the
scans were obtained from 50 to 700 °C at a heating rate of 10 °C/min.
   Tensile analysis of the blends and composites was carried out using universal testing machine (Instron, USA) (ASTM D 638). The
dumbbell-shaped samples were used for the analysis. A loading rate
of 10 mm/min was used, and the tests were carried out at room temperature. The tensile strength and modulus of the composites were
calculated using Win Test Analysis software. An average of five sample values was recorded. DHR-3 rheometer from TA Instruments was
used to characterize the rheological properties such as storage modulus (G') and complex viscosity elasticity (η^*). Disc shaped samples of
25 mm diameter and 2 mm thickness were measured at 190 °C using
a gap opening of 1.2 mm. Flow sweep experiment was carried out
apriori and all experiments were performed in linear viscoelastic region (LVR) at 0.5% strain from $\omega = 0.1$ to 100 rad/s.
   Wide angle X-ray diffraction (WAXRD) analysis of the materials
was performed with analytical powder diffractometer (X'Pert PRO)
using Cu Kα radiation ($\lambda = 1.5406$ Å) in reflection mode. A zero-background holder was used to minimize the noise. The samples were

step-scanned from 2 θ = 5-60° at room temperature using a step size of 2θ = 0.02° and a step time of 10 s. The microstructure of the surface of the nanocomposites was analyzed using scanning electron microscope (SEM) (FEI Quanta, FEG250, USA) with elemental mapping at accelerating voltages of 10-20 kV. The sample surfaces were sputter-coated with 3 nm thick gold layer.

Laser flash method (using Netzsch LFA 447) was employed to measure the thermal conductivity of the samples at varying temperatures (25, 50, 75 and 100 °C). The thermal conductivity was measured according to ASTM E1461, using the measured heat capacity and thermal diffusivity, with separately entered density data. The disc-shaped samples of 12 mm diameter and 2 mm thickness were used for the analysis and vespel was used as the standard. For the electrical conductivity analysis, four-point probe method containing a four-point resistivity probing fixture was employed.

9.3 Results and Discussion

9.3.1 Boron Nitride Composites

Figure 9.3 demonstrates the DSC melting and crystallization thermograms of PP, PP/PP-g-MA, PP/PA and the composites with BN. Table 9.1 also details the calorimetric properties of these materials. Addition of BN to PP without PP-g-MA resulted in a small decrease in the peak melting point, whereas the peak crystallization temperature was observed to slightly enhance as a function of BN content. The melting (and crystallization) enthalpy (normalized to 100% polymer content) was observed to increase with BN indicating nucleation effect as well as increased crystallinity of the polymer. Addition of PP-g-MA to PP resulted in a small reduction in the melt enthalpy. The addition of BN to the PP/PP-g-MA exhibited similar trend as the PP+BN composites. Addition of amorphous PA to PP resulted in a significant reduction in the melt enthalpy of the PP/PA blend. On incorporation 5% and 10% BN, peak melting and crystallization temperatures as well as melt enthalpy were observed to increase indicating that the filler resulted in enhanced crystallinity.

The thermal degradation behavior of the BN composites is demonstrated in Figure 9.4. Table 9.2 also reports the temperatures for 10 and 50% weight loss, along with char yield at 700 °C. Pure PP exhibited the temperatures for 10 and 50% weight loss as 404 and 448 °C, which were enhanced as a function of BN content in the composites.

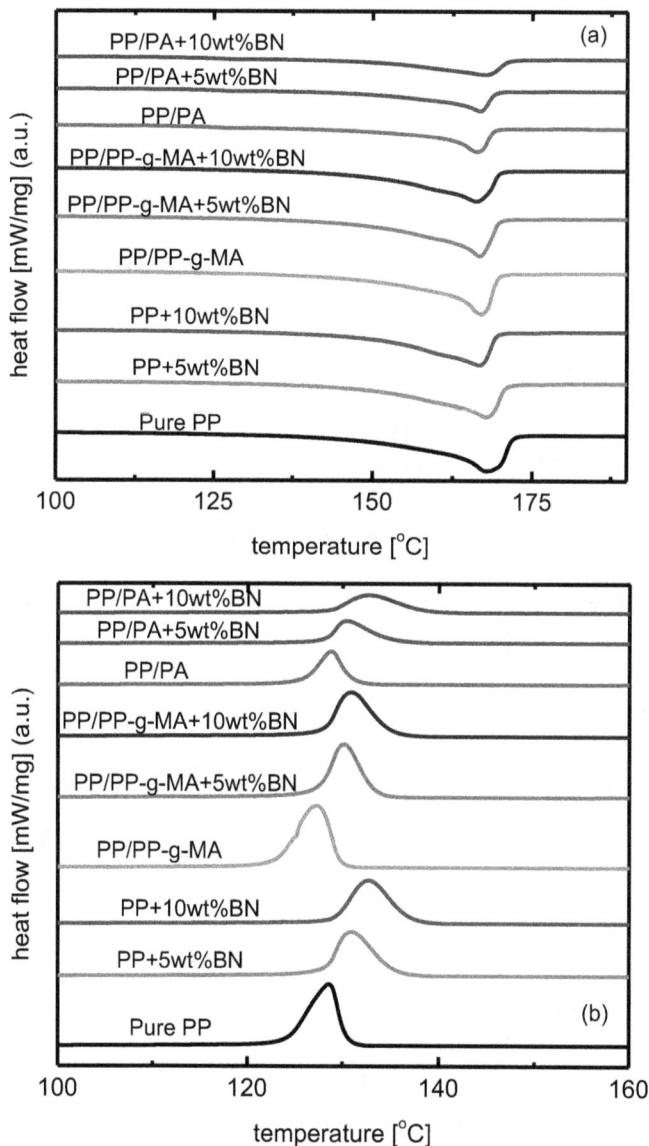

Figure 9.3 DSC (a) melting and (b) crystallization thermograms of BN composites.

The char yield also increased accordingly, with a value of 9.25% for the PP+10%BN composite as compared to 0.14% for pure PP. Addition of both PP-g-MA and PA to PP resulted in the enhancement in the initiation and peak degradation temperatures. Similar to the PP+BN

Table 9.1 Calorimetric properties of BN based nanocomposites

Sample	T_m (°C)	ΔH_m (J/g)	T_{crys} (°C)
Pure PP	168	95	129
PP+5wt%BN	168	101	130
PP+10wt%BN	166	102	132
PP/PP-g-MA	167	93	127
PP/PP-g-MA+5wt%BN	166	98	130
PP/PP-g-MA+10wt%BN	166	110	130
PP/PA	167	41	128
PP/PA+5wt%BN	167	45	130
PP/PA+10wt%BN	168	47	132

composites, the PP/PP-g-MA+BN and PP/PA+BN composites also exhibited an enhancement in the thermal stability as a function of BN content. Maximum extent of thermal stability improvement was observed in the case of PP/PA based blend composites. For instance, the PP/PA+10%BN composites exhibited $T_{10wt\%}$ and $T_{50wt\%}$ of 440 and 475 °C, in comparison to 410 and 459 °C for the PP/PA blend. The PP/PA based blend composites also exhibited a single step degradation pattern, indicating the compatibility of PA with PP. It is evident

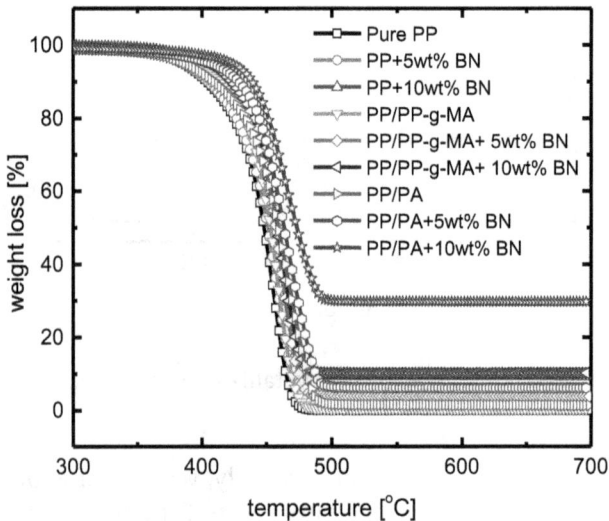

Figure 9.4 TGA thermograms of BN composites.

Table 9.2 Thermal stability and char yield of BN nanocomposites

Sample	$T_{10wt\%}$	$T_{50wt\%}$	Char yield @ 700 °C
Pure PP	404	448	0.14
PP+5wt%BN	406	452	4.80
PP+10wt%BN	424	458	9.25
PP/PP-g-MA	415	454	0.37
PP/PP-g-MA+5wt%BN	424	457	3.93
PP/PP-g-MA+10wt%BN	427	460	10.21
PP/PA	410	459	0.41
PP/PA+5wt%BN	430	465	6.16
PP/PA+10wt%BN	440	475	11.40

that BN phase in the composites formed a thermal barrier at the polymer-filler interface which delayed the release of thermal degradation products [12]. The weight loss as a function of time at a fixed temperature was also reduced in the composites, thus, confirming their superior thermal stability.

Figure 9.5 shows the X-ray diffractograms of the BN based composites. Pristine BN exhibited a sharp diffraction signal at 27°. The diffraction signal corresponding to BN was also observed in the PP+BN composites, indicating that the BN particles were not fully

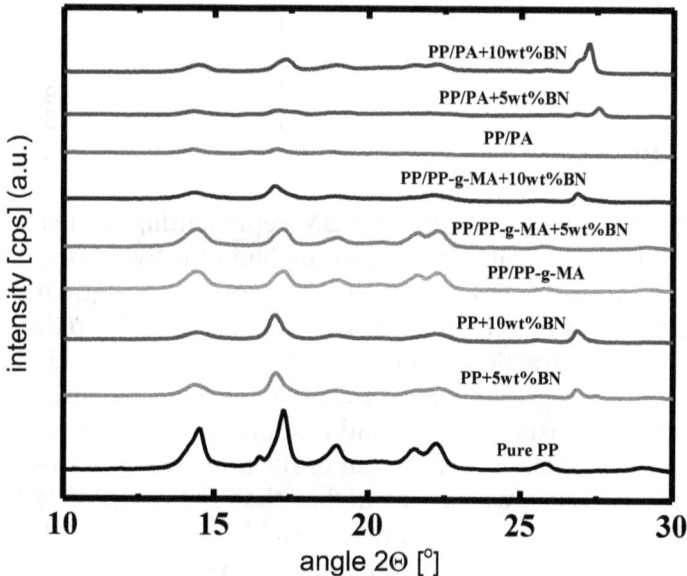

Figure 9.5 X-ray diffractograms of BN composites.

delaminated in the polymer. In the case of PP/PP-g-MA+BN composites, the BN signal was relatively reduced indicating higher degree of filler delamination due to the addition of the compatibilizer. The PP/PA based blend composites also exhibited a sharp diffraction peak corresponding to BN, the intensity of which was also dependent on the filler content. Overall, the crystalline phase of PP remained largely unaffected on the addition of BN, PP-g-MA and PA to PP. These observations indicated that enhancement of shear during the compounding process as well as increase in the compatibilizer content may be beneficial in achieving complete filler delamination.

The tensile properties of the BN based composites are reported in Table 9.3. PP exhibited a tensile modulus of 1044 MPa, which was

Table 9.3 Mechanical properties of BN based nanocomposites

Sample	UTS (MPa)	Modulus (MPa)	Extension at break (%)
Pure PP	33	1044	23
PP+5wt%BN	29	1051	12
PP+10wt%BN	21	1349	7
PP/PP-g-MA	32	956	7
PP/PP-g-MA+5wt%BN	29	1061	5
PP/PP-g-MA+10wt%BN	26	1423	5
PP/PA	36	1273	3
PP/PA+5wt%BN	25	1011	3
PP/PA+10wt%BN	24	853	3

enhanced to 1349 MPa with 10% BN, representing an increment of 30%. It indicated that the composites had effective stress transfer due to the presence of BN particles. The tensile strength and extension at break of the composites were observed to be reduced as a function of filler fraction. Addition of 5% PP-g-MA compatibilizer resulted in reduced extension indicating absence of plasticization. However, the tensile strength and modulus of PP were retained on addition of PP-g-MA. The addition of BN to PP/PP-g-MA resulted in enhanced modulus, which was higher than the composites without compatibilizer. This indicated that the addition of compatibilizer enhanced the extent of filler delamination, thereby, resulting in effective stress transfer. The observed changes in the polymer crystallinity on incorporating BN also resulted in corresponding effect on the tensile modulus. The reduction in tensile strength was also lesser in magnitude in the PP/PP-g-MA composites as compared to the composites without PP-g-MA. The PP/PA blend resulted in higher modulus as compared to PP due to the presence of stiffer PA. The PP/PA composite with 5% and 10% BN resulted in significant reduction in the modulus, probably due to the presence of filler aggregates. To further understand the factors leading to the observed behavior, it is needed to evaluate the morphology of the composites, especially phase distribution and filler localization.

Figure 9.6 presents the rheological performance of the BN composites as a function of angular frequency. As observed in Figure 9.6a, the materials exhibited an increase in the storage modulus with angular frequency. PP, PP/PP-g-MA and PP/PA exhibited similar trend for storage modulus as the tensile modulus, with PP/PA blend resulting in the highest storage modulus. The storage modulus of the composites was either same or lower than the corresponding polymer matrices. It indicated that the polymer melt was not restrained by the presence of low aspect ratio BN particles. Along with low aspect ratio, absence of strong interfacial interactions between the polymer and filler particles would have resulted in the observed phenomenon. PP+BN and PP/PP-g-MA+BN based composites also exhibited higher magnitude of loss modulus as compared to storage modulus, thus, indicating dominance of viscous response. As observed in Figure 9.6c, the PP/PA+BN composites exhibited highest value of viscosity. As expected, the viscosity of the materials decreased as a function of angular frequency.

The thermal conductivity performance of the composites is presented in Table 9.4 as a function of temperature. The increment in temperature resulted in a gradual decrease in the conductivity of the materials. At 25 °C, the thermal conductivity of PP was measured to be 0.225 W/mK, which was enhanced by more than 100% to a value of 0.545 W/mK on the addition of 5% BN. However, further increment in the BN content resulted in a decrease in the thermal conductivity of the composite, probably due to the presence of filler aggregates which resulted in a loss of conducting channels through the polymer. Addition of 5 wt% PP-g-MA further decreased the thermal conductivity to 0.173 W/mK, as compared to pure PP. However, the conductivity was observed to significantly increase with the addition of BN and the increment was also a function of BN content. For instance, the composite with 10% BN exhibited the thermal conductivity of 0.740 W/mK, which indicated an increment of >325%. It confirmed

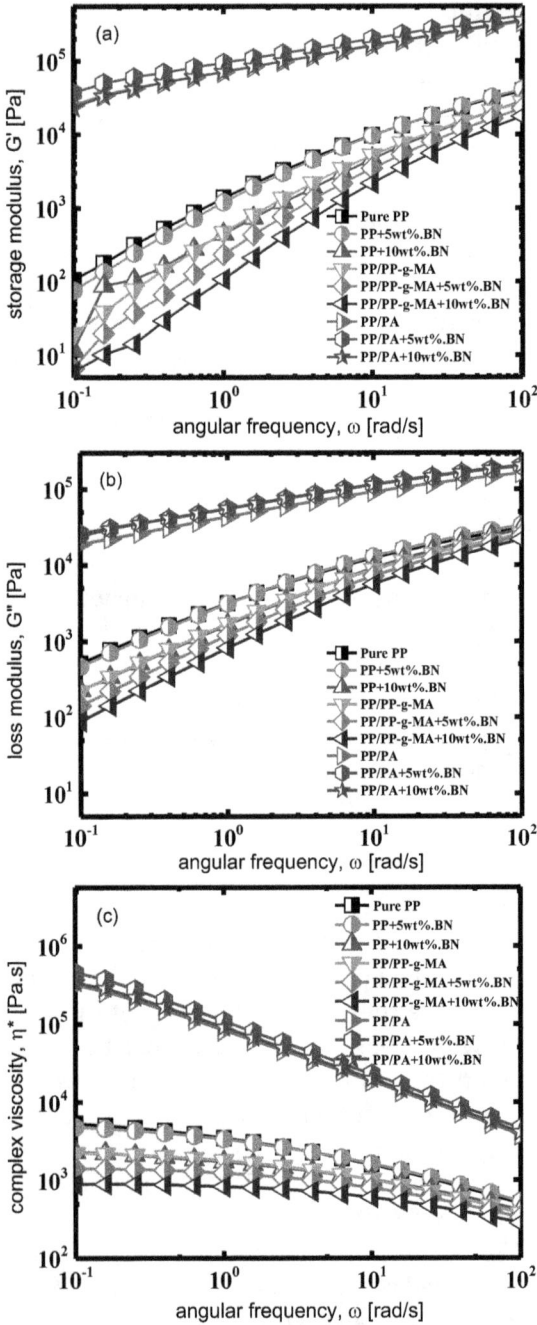

Figure 9.6 Storage modulus (a), loss modulus (b) and complex viscosity (c) of BN composites as a function of angular frequency.

that the addition of compatibilizer resulted in an effective filler dispersion, thereby leading to long-range thermally conducting paths. Blending PP with amorphous PA resulted in a significant reduction of thermal conductivity to 0.144 W/mK, as compared to pure PP. Similar to other cases, the addition of BN enhanced the thermal conductivity. The composite with 10% BN was observed to enhance the conductivity by >200%. Overall, the incorporation of BN was confirmed to be beneficial in significantly enhancing the thermal conductivity of the PP composites.

Table 9.4 Thermal conductivity of BN nanocomposites

Sample	Thermal conductivity (W/m K)			
	25 °C	**50 °C**	**75 °C**	**100 °C**
Pure PP	0.225	0.207	0.199	0.186
PP+5wt%BN	0.545	0.532	0.529	0.521
PP+10wt%BN	0.338	0.331	0.328	0.328
PP/PP-g-MA	0.173	0.168	0.168	0.165
PP/PP-g-MA+5wt%BN	0.688	0.678	0.672	0.669
PP/PP-g-MA+10wt%BN	0.740	0.738	0.729	0.721
PP/PA	0.144	0.144	0.138	0.128
PP/PA+5wt%BN	0.413	0.398	0.389	0.375
PP/PA+10wt%BN	0.435	0.431	0.428	0.422

The surface electrical resistivity of the BN based composites is presented in Table 9.5. Neat PP, PP/PP-g-MA and PP/PA displayed the surface resistivity of about 10^9 Ωm. The surface resistivity of PP remained unchanged by the incorporation of 5% and 10% BN. In the case of composites generated in the presence of PP-g-MA, the surface resistivity was observed to increase by an order of magnitude as compared to PP/PP-g-MA. As the surface resistivity is directly proportional to the insulating property of the composites, it indicated an enhancement of the electrical insulation of the materials. It is well known that the BN particle exhibit an electrical insulation property [13]. Thus, the addition of BN particles developed electrical insulating properties in the polymer matrix. Though the presence of compatibilizer led to enhancement in the surface resistivity due to enhanced filler dispersion, however, the resistivity was not affected by the filler fraction. Similarly, in the case of PP/PA composites, the resistivity was observed to increase by an order of magnitude as compared to the PP/PA matrix, but remained unchanged with the filler fraction.

Table 9.5 Electrical resistivity of BN nanocomposites

Sample	Electrical resistivity (Ωm)
Pure PP	2.0×10^9
PP+5wt%BN	2.0×10^9
PP+10wt%BN	2.1×10^9
PP/PP-g-MA	3.7×10^9
PP/PP-g-MA+5wt%BN	4.9×10^{10}
PP/PP-g-MA+10wt%BN	4.5×10^{10}
PP/PA	3.0×10^9
PP/PA+5wt%BN	3.0×10^{10}
PP/PA+10wt%BN	3.0×10^{10}

9.3.2 Graphene Composites

Table 9.6 demonstrates the calorimetric properties of graphene based composites. Figure 9.7 also shows the melting and crystallization thermograms of the nanocomposites. In the case of PP+G composites, no significant change in the peak melting and crystallization temperatures of PP was observed on the addition of graphene nanosheets. However, the melt enthalpy was observed to slightly increase as a function of graphene content, thus, indicating an increment in the polymer crystallinity [14]. Similarly, in the presence of the compatibilizer, the crystallinity of the nanocomposites was also observed to increase to a small extent, while the peak melting and crystallization

Table 9.6 Calorimetric properties of graphene based nanocomposites

Sample	T_m (°C)	ΔH_m (J/g)	T_{crys} (°C)
Pure PP	168	95	129
PP+3wt%G	167	95	128
PP+5wt%G	167	99	128
PP/PP-g-MA	167	93	127
PP/PP-g-MA+3wt%G	167	94	126
PP/PP-g-MA+5wt%G	167	95	126
PP/PA	167	41	128
PP/PA+3wt%G	168	52	127
PP/PA+5wt%G	168	59	128

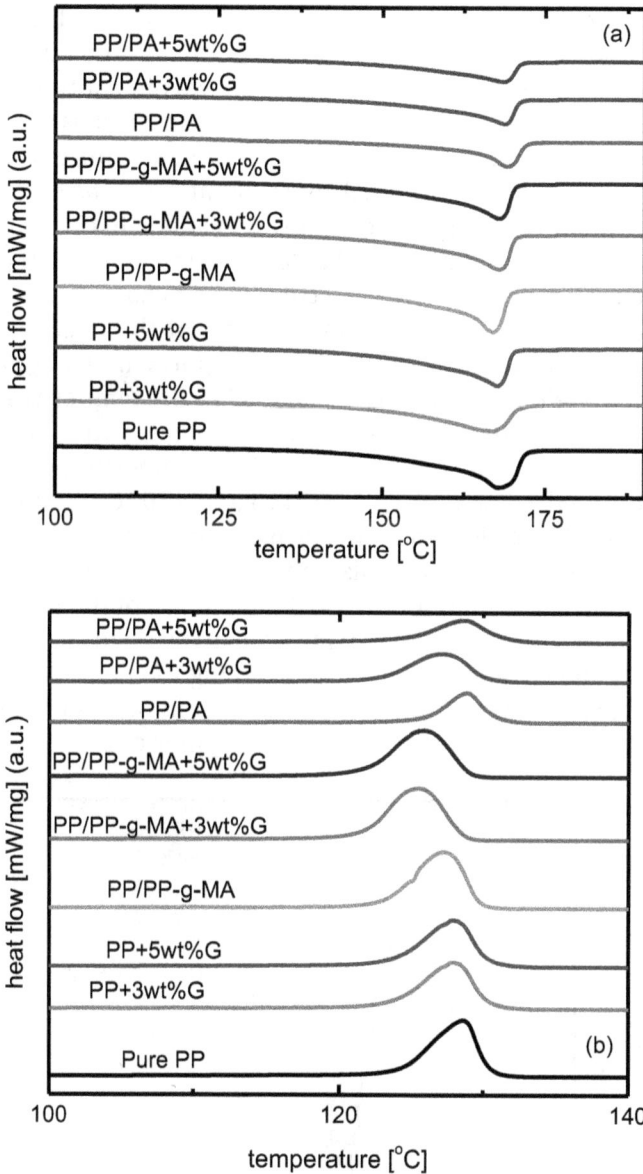

Figure 9.7 DSC (a) melting and (b) crystallization thermograms of graphene based composites.

temperatures remained constant. The composites with PP/PA also exhibited a significant increment in the melt enthalpy. For instance, as compared to 41 J/g for PP/PA, the composites with 5% graphene exhibited the melt enthalpy of 59 J/g. Overall, the addition of graphene impacted the polymer crystallinity, however, no nucleation effect was observed. The observed behavior was in contrast with the BN composites, which exhibited both nucleation effect as well as crystallinity enhancement on incorporating BN. In addition, graphene impacted the crystallinity of PP/PA more significantly as compared to BN, even though a larger fraction of BN was added in the composites. Larger aspect ratio of graphene platelets as compared to BN particles may have resulted in the observed composite behaviors.

Figure 9.8 shows the TGA thermograms of the nanocomposites with graphene. In addition, Table 9.7 also reports the $T_{10wt\%}$, $T_{50wt\%}$ as well as char yield of the materials. Similar to BN composites, the addition of graphene to PP resulted in the enhancement of both $T_{10wt\%}$ and $T_{50wt\%}$ temperatures. For instance, the composite with 5% graphene exhibited $T_{10wt\%}$ and $T_{50wt\%}$ of 440 and 470 °C, as compared to 404 and 448 °C for PP. In the presence of the compatibilizer and PA, the addition of graphene was also observed to increase the thermal stability and char yield of PP. Thus, the presence of the graphene increased the rigidity of the polymeric matrix resulting in the restriction to the chain mobility, thus, delaying the degradation of the polymer [15]. Furthermore, the development in the thermal stability can also be attributed to the graphene-polymer chains interactions,

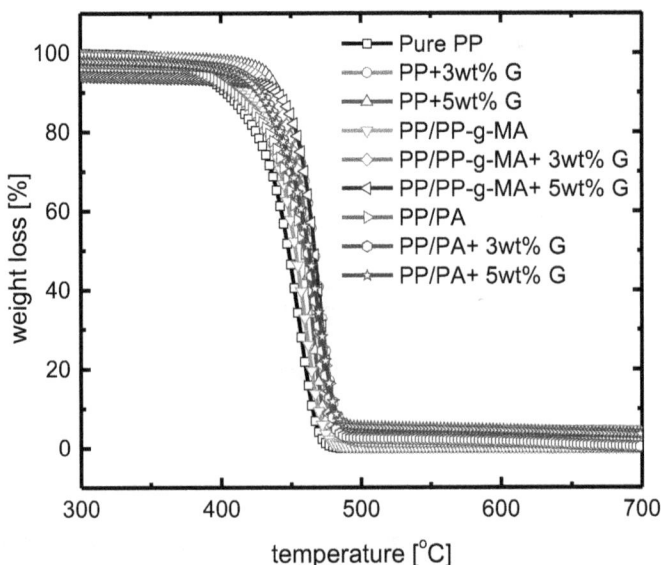

Figure 9.8 TGA thermograms of nanocomposites with graphene.

which lead to the formation of network-structured layer of graphene in the polymer matrix [14]. The composites also exhibited delayed degradation when tested under isothermal conditions at 190 °C, as a function of time. It, thus, confirmed that the composites have suitable thermal stability at the processing temperature used to generate the nanocomposites.

Table 9.7 Thermal stability and char yield of graphene nanocomposites

Sample	$T_{10wt\%}$	$T_{50wt\%}$	Char yield @ 700 °C
Pure PP	404	448	0.14
PP+3wt%G	437	466	4.16
PP+5wt%G	440	470	5.58
PP/PP-g-MA	415	454	0.37
PP/PP-g-MA+3wt%G	436	462	3.24
PP/PP-g-MA+5wt%G	440	466	3.32
PP/PA	410	459	0.41
PP/PA+3wt%G	428	461	1.40
PP/PA+5wt%G	430	462	3.53

The X-ray diffractograms of graphene nanocomposites are demonstrated in Figure 9.9. Graphene was observed to have the

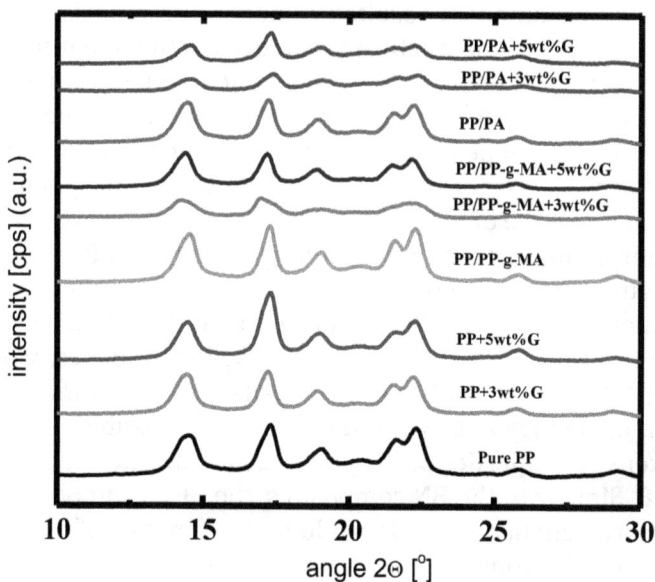

Figure 9.9 X-ray diffractograms of graphene nanocomposites.

characteristic diffraction peak at $2\theta \approx 26.5°$ (not shown). In the PP+G composites, the graphene characteristic peak was observed to be absent, thus, indicating a good degree of filler dispersion. PP exhibited crystalline diffraction peaks corresponding to (1 1 0), (0 4 0), (1 3 0), (1 1 1), (0 4 1), (0 6 0) and (2 2 0) planes (α monoclinic), which were observed to remain unchanged with the addition of graphene, thus, indicating no change in the PP crystalline structure. Similarly, the crystal structure of PP phase was also retained in both PP/PP-g-MA and PP/PA based composites. Diffraction peak corresponding to graphene was also absent in these composites due to effective filler delamination, even at 5 wt% filler content [16,17]. The observed behavior was in contrast with the BN composites, where the processing conditions employed for the generation of nanocomposites did not lead to effective filler delamination, and the filler characteristic diffraction peak was observed in the composite diffractograms.

Morphology of the fracture surface of the graphene composites was also studied with scanning electron microscopy, as demonstrated in Figure 9.10 for PP, PP/PP-g-MA and PP/PA composites with 5 wt% graphene. In the case of PP+G composite (Figure 9.10a), brittle failure was observed with significant extent of micro-voids present on the surface. Lack of adhesion between the polymer-filler phases may have resulted in such morphology. The addition of PP-g-MA to PP in the composite resulted in more optimal surface morphology (Figure 9.10b). The fracture was more ductile in nature, and the micro-voids were significantly reduced due to effective filler dispersion as well as enhanced polymer-filler interaction. The exfoliated graphene layers can be envisaged to absorb more rupture energy of the polymer matrix and reduce the crack formation phenomenon. The fracture surface of PP/PA+G composite exhibited completely different morphology as the corresponding PP and PP/PP-g-MA composites (Figure 9.10c). The fracture was also ductile in nature, though, the composite did not exhibit significant extension. Similar observations were also observed for the BN based composites, where the PP/PP-g-MA and PP/PA based composites exhibited more optimal surface morphology as compared to the PP composites.

The tensile properties of the graphene composites are detailed in Table 9.8. Similar to the BN composites, the PP+G composites exhibited an increment in the tensile modulus as a function of filler content. For instance, the composite with 5 wt% graphene resulted in an increase of 46% in the modulus, as compared to PP. In addition, the extent of modulus enhancement in the case of PP+G composites was

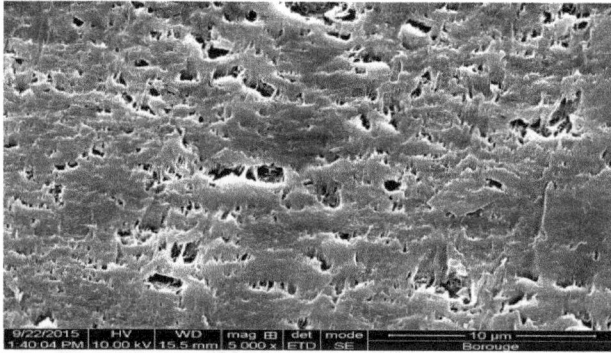

(a)

(b)

(c)

Figure 9.10 Scanning electron micrographs of the fracture surfaces of (a) PP+5wt%G, (b)PP/PP-g-MA+5wt%G, and (c) PP/PA+5wt%G nanocomposites.

Table 9.8 Mechanical properties of graphene based nanocomposites

Sample	UTS (MPa)	Modulus (MPa)	Extension at break (%)
Pure PP	33	1044	23
PP+3wt%G	29	1151	5
PP+5wt%G	22	1522	3
PP/PP-g-MA	32	956	7
PP/PP-g-MA+3wt%G	23	1488	3
PP/PP-g-MA+5wt%G	21	1903	2
PP/PA	36	1273	3
PP/PA+3wt%G	24	1852	2
PP/PA+5wt%G	23	2254	2

also significantly superior to the corresponding PP+BN composites. The tensile strength and extension of the PP+G composites were observed to be negatively impacted on the addition of graphene. The addition of the PP-g-MA to PP for the composite generation resulted in further increase in the modulus of the composites, as compared to PP+G composites. The PP/PP-g-MA composite with 5 wt% graphene exhibited a modulus of 1903 MPa, which indicated an increment of ~100%. Thus, the addition of compatibilizer resulted in enhanced polymer-filler interfacial interactions and higher extent of filler delamination, thereby, leading to effective stress transfer. Similarly, the PP/PA composites also had a significant enhancement in the tensile modulus on graphene incorporation. In comparison to 1273 MPa for PP/PA blend, the tensile modulus of the composite with 5 wt% graphene was measured to be 2254 MPa. Similar to the PP and PP/PP-g-MA composites, the strength and extension at break of the PP/PA composites were observed to reduce on graphene incorporation. Overall, the comparison among the graphene and BN composites revealed much higher extent of modulus enhancement in the case of graphene as reinforcing filler.

Figure 9.11 demonstrates the storage and loss modulus as well as complex viscosity of graphene composites as a function of angular frequency. As shown in Figure 9.11a, the addition of graphene to PP resulted in a significant enhancement in storage modulus, which was also a function of filler content. In addition, the storage modulus curves for the composites exhibited nearly frequency independent behavior. Similarly, the PP/PP-g-MA based composites also exhibited

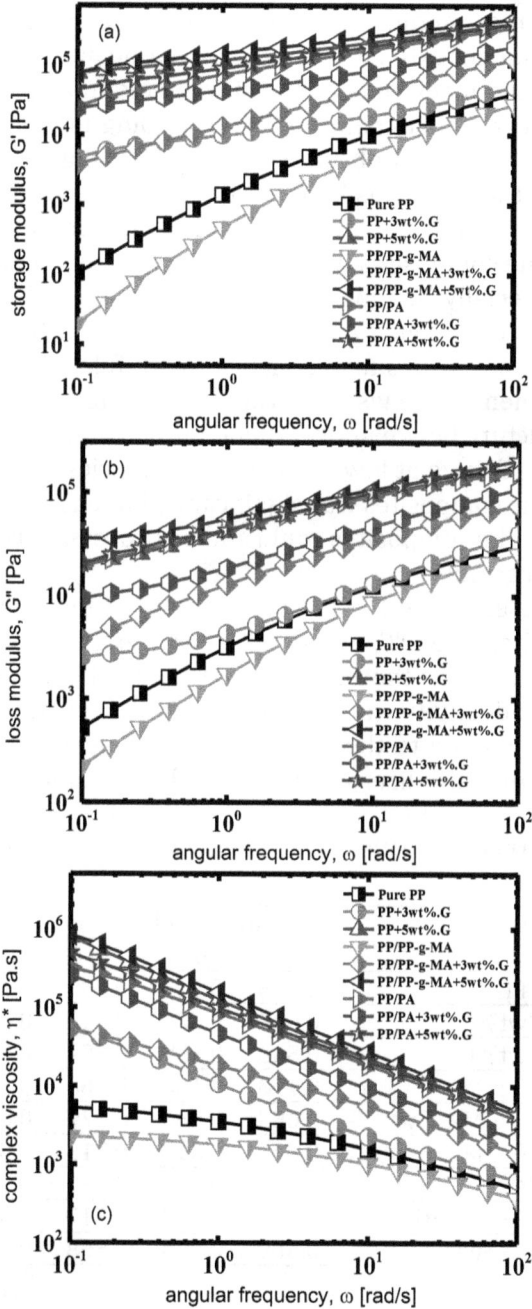

Figure 9.11 Storage modulus (a), loss modulus (b) and complex viscosity (c) of graphene nanocomposites as a function of angular frequency.

enhanced storage modulus over the whole angular frequency range. Owing to the higher extent of filler delamination due to polymer-filler interactions, the magnitude of the storage modulus of PP/PP-g-MA composites was higher than the corresponding PP composites. The modulus of PP/PA was observed to be much higher than PP and PP/PP-g-MA, though it was less affected by the incorporation of graphene as compared to the PP and PP/PP-g-MA composites. The loss modulus of the composites was lower in magnitude than the storage modulus, especially at lower angular frequencies, thus, indicating dominant elastic behavior. As shown in Figure 9.11c, the complex viscosity of the composites also enhanced with filler fraction. The observed increment in the viscosity can be attributed to the interlinked network structure between the graphene and polymer matrix, which is especially effective at lower angular frequencies. Even though the viscosity of the composites was significantly increased on incorporating graphene, the composites could still be processed using the same processing protocols.

Table 9.9 presents the thermal conductivity of the graphene based nanocomposites. At 25 °C, the addition of 3% graphene to PP resulted in an increment in the thermal conductivity by 180%. A slight reduction in the conductivity was observed in the composite with 5% graphene, probably due to the presence of filler aggregates in the absence of the compatibilizer which hinder the formation of conducting

Table 9.9 Thermal conductivity of graphene nanocomposites

Sample	Thermal conductivity (W/m K)			
	25 °C	50 °C	75 °C	100 °C
Pure PP	0.225	0.207	0.199	0.186
PP+3wt%G	0.632	0.628	0.619	0.620
PP+5wt%G	0.566	0.559	0.552	0.552
PP/PP-g-MA	0.173	0.168	0.168	0.165
PP/PP-g-MA+3wt%G	0.683	0.680	0.677	0.672
PP/PP-g-MA+5wt%G	0.821	0.818	0.811	0.807
PP/PA	0.144	0.144	0.138	0.128
PP/PA+3wt%G	0.525	0.520	0.517	0.515
PP/PA+5wt%G	0.568	0.563	0.558	0.553

channels in the polymer matrix. The effect of compatibilizer in enhancing the filler dispersion was evident from the higher magnitude of thermal conductivity in the PP/PP-g-MA composites, which was also a function of the filler content. For instance, the composite with 5% graphene was observed to have 380% increase in conductivity as compared to PP/PP-g-MA. Similar increase was observed in the case of PP/PA composites, where the composite with 5% graphene exhibited thermal conductivity of 0.568 W/mK as compared to 0.144 W/mK for PP/PA. Overall, the thermal conductivity of the composites was significantly improved on the addition of a small amount of graphene. In addition, the graphene composites also exhibited much higher extent of increase in the thermal conductivity as compared to the BN composites.

The surface resistivity of the graphene reinforced nanocomposites is presented in Table 9.10. Addition of graphene was observed to reduce the surface resistivity of PP by four orders of magnitude. As the surface resistivity is inversely proportional to the surface conductivity, it indicated a significant increase in the electrical conductivity.

Table 9.10 Electrical resistivity of graphene nanocomposites

Sample	Electrical resistivity (Ωm)
Pure PP	2.0×10^9
PP+3wt%G	3.0×10^5
PP+5wt%G	4.6×10^5
PP/PP-g-MA	3.7×10^9
PP/PP-g-MA+3wt%G	4.6×10^4
PP/PP-g-MA+5wt%G	1.4×10^5
PP/PA	3.0×10^9
PP/PA+3wt%G	4.8×10^4
PP/PA+5wt%G	6.4×10^6

Thus, the incorporation of graphene in the polymer increased the mobility of the conducting network and the density of the potential charge carriers [15]. PP/PP-g-MA and PP/PA composites with 3 wt% graphene also resulted in five orders of magnitude reduction in the surface resistivity. The composites with 5 wt% graphene had a small extent of increase in the surface resistivity, however, it was still 3-4 orders of magnitude lower than respective polymer matrices, thus, indicating enhanced electrical conductivity of the composites.

9.4 Conclusions

In the current study, mechanical, thermal, rheological and conductivity properties of PP, PP/PP-g-MA, PP/PA nanocomposites, prepared with graphene and boron nitride filler particles, were studied. The nanocomposites exhibited optimal filler dispersion in the presence of PP-g-MA, though the extent of dispersion of graphene was observed to be superior than BN particles. The filler particles enhanced the crystallinity as well as thermal stability of the polymer matrices. Furthermore, the tensile and elastic moduli of graphene composites were much higher in magnitude than the corresponding BN composites. Addition of 5 wt% graphene to the polymers resulted in significant increment in the thermal conductivity. 4-5 order of magnitude reduction in the surface resistivity of the composites was observed on the addition of 3 wt% graphene. On the other hand, the BN composites resulted in an enhancement of the surface resistivity of the composites. The developed nanocomposite materials represent advanced materials with high potential of application in the areas such as heat exchangers, electronic packaging, etc.

References

1. Pavlidoua, S., and Papaspyrides, C. D. (2008) A review on polymer–layered silicate nanocomposites. *Progress in Polymer Science*, **33**, 1119-1198.
2. Mittal, V. (2007) Polypropylene-layered silicate nanocomposites: Filler matrix interactions and mechanical properties. *Journal of Thermoplastic Composite Materials*, **20**, 575-599.
3. Mukhopadhyay, P., and Gupta, R. K. (2011) Trends and frontiers in graphene-based polymer nanocomposites. *Plastics Engineering*, **67**, 32-42.
4. Mittal, V. (2014) Functional polymer nanocomposites with graphene: A review. *Macromolecular Materials and Engineering*, **299**, 906-931.
5. Pawar, S., Mittal, V., and Matsko, N. (2017) PP/EVA-MA blend nanocomposites with graphene nano-sheets. In: *Functional Polymer Blends and Nanocomposites*, Mittal, V. (ed.), Central West Publishing. Australia, pp. 107-136.
6. Mittal, V., Pannirselvam, M., and Griffin, G. (2017) Polypropylene/polystyrene blend nanocomposites with graphene. In: *Functional Polymer Blends and Nanocomposites*, Mittal, V. (ed.), Central West Publishing. Australia, pp. 137-164.
7. Chen, H., Ginzburg, V. V., Yang, J., Yang, Y., Liu, W., Huanf, Y., Du, L., and Chen, B. (2016) Thermal conductivity of polymer-based composites: Fundamentals and applications. *Progress in Polymer Science*, **59**, 41-85.

8. Tavman, I. H., and Akinci, H. (2000) Transverse thermal conductivity of fiber reinforced polymer composites. *International Communications in Heat and Mass Transfer*, **27**(2), 253-261.

9. Kusunose, T., Yagi, T., Firoz, S. H., and Sekino, T. (2013) Fabrication of epoxy/silicon nitride nanowire composites and evaluation of their thermal conductivity. *Journal of Materials Chemistry A*, **1**(10), 3440-3445.

10. Yu, J., Chen, Y., Wuhrer, R., Liu, Z., and Ringer, S. P. (2005) In situ formation of BN nanotubes during nitriding reactions. *Chemistry of Materials*, **17**(20), 5172-5176.

11. Hill, R. F., and Supancic, P. H. (2002) Thermal conductivity of platelet-filled polymer composites. *Journal of the American Ceramic Society*, **85**(4), 851-857.

12. Li, T., and Hsu, S. L.-C. (2011) Preparation and properties of thermally conductive photosensitive polyimide/boron nitride nanocomposites. *Journal of Applied Polymer Science*, **121**, 916-922.

13. Hou, J., Li, G., Yang, N., Qin, L., Grami, M. E., Zhang, Q., Wang, N., and Qu, X. (2014) Preparation and characterization of surface modified boron nitride epoxy composites with enhanced thermal conductivity. *RSC Advances*, **4**, 44282-44290.

14. Zhao, S., Chen, F., Zhao, C., Huang, Y., Dong, J. Y., and Han, C. C. (2013) Interpenetrating network formation in isotactic polypropylene/graphene composites. *Polymer*, **54**(14), 3680-3690.

15. Fim, F. d. C., Basso, N. R. S., Graebin, A. P., Azambuja, D. S., and Galland, G. B. (2013) Thermal, electrical, and mechanical properties of polyethylene–graphene nanocomposites obtained by in situ polymerization. *Journal of Applied Polymer Science*, **128**(5), 2630-2637.

16. Qiu, F., Hao, Y., Li, X., Wang, B., and Wang, M. (2015) Functionalized graphene sheets filled isotactic polypropylene nanocomposites. *Composites, Part B: Engineering*, **71**, 175-183.

17. Milani, M. A., Gonzalez, D., Quijada, R., Basso, N. R. S., Cerrada, M. L., Azambuja, D. S., and Galland, G. B. (2013) Polypropylene/graphene nanosheet nanocomposites by in situ polymerization: Synthesis, characterization and fundamental properties. *Composites Science and Technology*, **84**, 1-7.

10

Polyethylene-Thermally Reduced Graphene Nanocomposites: Comparison of Masterbatch and Direct Melt Mixing Approaches on Mechanical, Thermal, Rheological and Morphological Properties

10.1 Introduction

Nanocomposites represent a new class of materials having superior mechanical, electrical, thermal and gas barrier properties along with lightweight and optical transparency [1-7]. Various 1-D and 2-D nanofillers (e.g. fibrous and platy fillers) due to their high aspect ratio and geometrical shape have the potential to influence the polymer properties at very low fractions. However, to achieve superior performance especially at lower filler fractions, dispersion of the reinforcement phase at nanoscale is of utmost importance [1,6]. Thermodynamic factors such as interfacial compatibility of polymer and filler phases as well as kinetic factors such as filler shape and size, dispersion techniques and equipment, time of mixing and applied shear lead to final morphology in the polymer nanocomposites [8-13]. Absence of any interfacial interactions between the polymer and filler phases and mismatch of polarity between the low surface energy polymers, especially polyolefins, and high surface energy reinforcements often leads to non-optimum filler dispersion, thus, resulting in performance similar to conventional macro-composites.

The development of free-standing single-layer graphene from inexpensive sources (graphite) has opened the gateway for novel functional polymer nanocomposites with much superior properties as compared to property increments usually achieved from other conventional fillers [14-20]. Graphene sheets are one-atom-thick two-dimensional layers of sp^2-bonded carbon and can effectively enhance electrical and thermal conductivity, gas impermeability and mechanical properties of polymers [14-20]. For instance, a single

Ali U. Chaudhry and Vikas Mittal**, The Petroleum Institute (part of Khalifa University of Science and Technology), Abu Dhabi, UAE*
*Current address: Texas A&M University, Qatar; **Current address: Bletchington, Wellington County, Australia
© 2019 Central West Publishing, Australia

defect-free graphene layer has been estimated to have Young's modulus of ≈1.0 TPa, intrinsic strength ≈42 N/m, thermal conductivity ≈4840-5300 W/(m.K), electron mobility exceeding 25,000 cm^2/V.s, excellent gas impermeability and specific surface area of ≈2630 m^2/g [21]. Thus, even with some degree of defects in the structure, graphene platelets represent functional reinforcement material to achieve nanocomposites with exceptional properties. It is also confirmed by large number of recent studies reported on various polymer-graphene nanocomposites systems [21-26].

Significant research efforts have been dedicated to enhance the dispersion of nanofillers by using different mixing techniques, modification of polymer backbone or filler surfaces, use of compatibilizers (functional polymers) and coupling agents, etc. [27]. Thermally reduced graphene as compared to graphene oxide (GO) lacks the presence of functional groups and has strong interlayer cohesive energy and surface inertia, which make it less compatible with polymers for generating nanocomposites [28]. In order to improve the dispersion of thermally reduced graphene in polymer matrices, different routes have been adopted, such as surface modification by phenyl isocyanate and porphyrin [29] as well as use of surfactants and compatibilizers for exfoliation and dispersion [30,31]. Fang *et al.* [29] covalently linked graphene with polystyrene using atom transfer radical polymerization and showed improvement in thermal and mechanical properties. Castelain *et al.* [32] reported different chemical routes to functionalize graphene with short-chain polyethylene which influenced the mechanical properties of graphene-based high density polyethylene (HDPE) nanocomposites. Vasileiou *et al.* [33] reported the dispersion and properties of polyethylene-graphene composite using maleated linear low-density polyethylene (LLDPE) derivatives and thermally reduced graphene oxide through a non-covalent compatibilization approach. Yu *et al.* [34] prepared a flame retardant functionalized GO (FRs-FGO) using amine functional GO and phosphoramide oligomer, which was incorporated into PP along with PP-grafted maleic anhydride (PP-g-MA). In an interesting study, Seo *et al.* [35] also reported the compatibility of functionalized graphene with polyethylene as well as its copolymers as a function of molecular weight and polarity.

Many studies have also focused on different methods of nanocomposite generation to achieve efficient filler dispersion. Wang *et al.* [36] prepared low density polyethylene (LDPE)/graphene nanocomposites using vinyl functionalized graphene sheets and LDPE through solution blending method. In a recent study, different methods of dispersion for graphene in LLDPE were adopted, and their effect on mechanical properties was reported. The methods included solution mixing and melt mixing by co-counter-rotating and modified co-rotating screw systems [37]. It was observed that solution mixing resulted in better properties than co-counter-rotating and modified co-rotating methods [37]. Giannelis *et al.* [38] incorporated functionalized graphene sheets in poly(vinylidene fluoride) matrix by solution mixing and melt blending which resulted in percolation at a much lower filler amount. As the high molecular weight polymers, especially polyolefins, are not soluble in common solvents, solution mixing method is, thus, less effective. As an alternative, solvent based dispersion of fillers in compatibilizers or functional polymers (solution mixing) is a simple and effective processing route for the fabrication of hybrid masterbatches. In this method, the filler and the functional polymer interact through relatively weak dispersive forces [31,39]. Subsequently, the incorporation of masterbatch by melt mixing with high molecular weight polymer matrices can lead to a better state of filler dispersion than directly melt mixed polymer, compatibilizer and filler. In a recent study, HDPE nanocomposites were generated by melt mixing with chlorinated polyethylene-graphene masterbatch [40], however, more efforts are needed to establish direct comparison between the masterbatch and direct melt mixing methods.

The goal of the current study was to explore the effectiveness of the masterbatch approach over direct melt mixing method in improving the filler dispersion and resulting composite properties. Compatibilizers of different chemical architectures and polarities were chosen. The amount of the composite constituents was held constant in order to correlate the observed property changes to the method of nanocomposite generation and specific compatibilizer.

10.2 Experimental

10.2.1 Materials

High density polyethylene FB1460 was supplied by Abu Dhabi Polymers Company Limited, UAE. It was in the form of white pellets with specific gravity 0.946 and peak melting point ~131 °C (supplier data). The melt flow index (MFI) for the polymer at 2.16 kg and 190 °C was reported by the supplier as <0.1 g/10 min, indicating high molecular weight. Three copolymer based compatibilizers (ethylene acrylic acid copolymer (EAA), ethylene vinyl acetate-*graft*-maleic anhydride (EVA-MA) and ethylene methacrylic acid copolymer with zinc ion (EMAZ)) of different specifications and functionalities were procured from suppliers as mentioned in Table 10.1. The polymer materials were used as obtained. Thermally reduced graphene (G) was prepared through thermal exfoliation of precursor graphite oxide [41] using modified Hummer's method [42], as reported earlier [40,43]. The confirmation of graphene formation was achieved through Raman spectroscopy, microscopy and X-ray diffraction. The average size of the graphene platelets was 2 μm, whereas the average thickness was 10 nm (few multilayers).

Table 10.1 Details of the polymer, compatibilizers, and filler used for the study

Name	Repeat Unit	Specifications
High density polyethylene (HDPE)	$\left[CH_2-CH_2- \right]_n$	Trade name: Borouge FB1460 MFI (2.16 kg, 190 °C): <0.1 g/10 min Density: 0.946 g/cm^3 Peak melting temperature: 131 °C
Ethylene acrylic acid copolymer (EAA)	$\left[CH_2-CH_2 \right]_x \left[CH-CH_2 \right]_y$ (with $O{=}C{-}OH$ pendant)	Trade name: Exxon Mobil ESCOR 5050 Co-monomer content: 9% AA MFI (2.16 kg, 190 °C): 8.4 g/10 min Density: 0.936 g/cm^3 Peak melting temperature: 97 °C
Ethylene vinyl acetate-g-maleic anhydride (EVA-MA)	$\left[CH_2-CH_2 \right]_n \left[CH-CH_2 \right]_n$ (with maleic anhydride pendant)	Trade name: Dupont Fusabond C190 Co-monomer content: 12-15% MA MFI (2.16 kg, 190 °C): 16 g/10 min Density: 0.950 g/cm^3 Peak melting temperature: 71 °C

Ethylene methacrylic acid copolymer with zinc ion (EMAZ)		Trade name: Dupont Surlyn 9020 Co-monomer content: 14% MFI (2.16 kg, 190 °C): 1 g/10 min Density: 0.960 g/cm^3 Peak melting temperature: 85 °C
Thermally reduced graphene (G)		Obtained using modified Hummer's method C and O fractions: 77.5% and 18.4% respectively Density: 0.0161 g/cm^3

10.2.2 Generation of Nanocomposites

The nanocomposites were generated either by masterbatch approach (solution mixing followed by melt mixing) or direct melt mixing method, as shown in Scheme 10.1. For the masterbatch method, solution of compatibilizer (functional polymer (FP)) in *p*-xylene was generated by stirring at 100 °C for 2 h (3% solid content). Suspension of graphene in *p*-xylene was obtained by stirring at room temperature for 1 h followed by sonication for 10 min. The two components were then mixed at room temperature, sonicated and gradually mixed at 100 °C for further 1 h. The blend was stirred and gradually brought to room temperature. The solution was kept overnight under vacuum at room temperature followed by vacuum drying at 40 °C. Two masterbatches with graphene concentration of 9% and 17% were obtained. The so-obtained masterbatches of compatibilizer with graphene were subsequently melt mixed in the extruder with appropriate amount of HDPE to generate nanocomposite with 1% filler content and compatibilizer content 5% (using masterbatch with 17% graphene concentration) or 10% ((using masterbatch with 9% graphene concentration). In the direct melt mixing method, all three components of the composite system were directly mixed in the extruder to generate nanocomposites. HDPE/G composite was also generated similarly by melt mixing. For notation

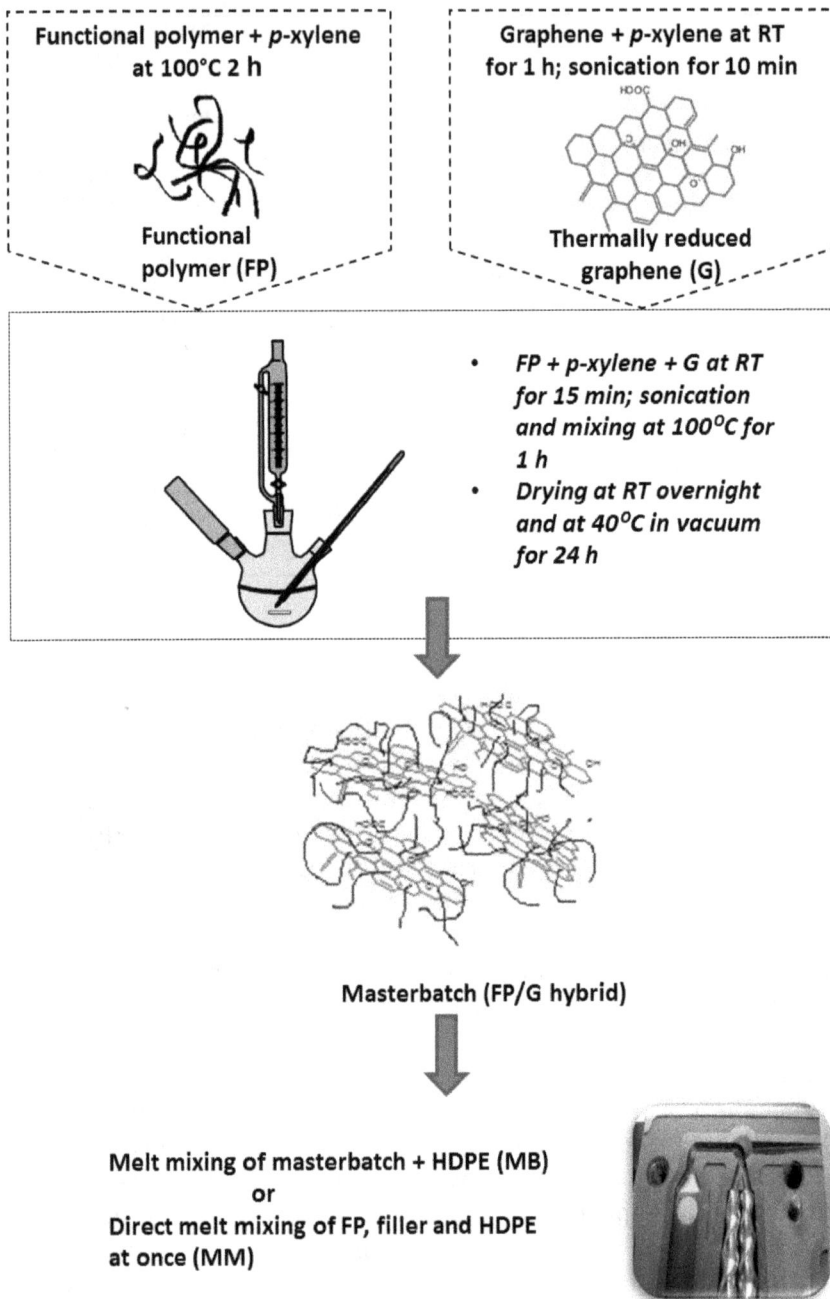

| Functional polymer + *p*-xylene at 100°C 2 h | Graphene + *p*-xylene at RT for 1 h; sonication for 10 min |

Functional polymer (FP)

Thermally reduced graphene (G)

- *FP + p-xylene + G at RT for 15 min; sonication and mixing at 100°C for 1 h*
- *Drying at RT overnight and at 40°C in vacuum for 24 h*

Masterbatch (FP/G hybrid)

Melt mixing of masterbatch + HDPE (MB)
or
Direct melt mixing of FP, filler and HDPE
at once (MM)

Scheme 10.1 Schematic representation of the generation of nanocomposites using masterbatch and direct melt mixing approaches.

purposes, the technique using masterbatch approach was denoted as 'MB', whereas the direct melt mixing was named as 'MM'. Melt mixing was carried out using a mini twin screw extruder (MiniLab HAAKE Rheomex CTW5, Germany), which had a screw length and screw diameter of 109.5 mm and 5/14 mm conical respectively. Batch size of 4.5 g was used and the shear mixing was performed for 6 min at 60 rpm. The compounding temperature used for nanocomposite generation was 190 °C. Pure polymer was also processed similarly by subjecting it to similar shear and thermal conditions. Disc and dumbbell shaped test specimens were injection molded using a mini injection molding machine (HAAKE MiniJet, Germany) using same processing temperature as used during compounding. The injection pressure was 700 bars for 6 s, whereas holding pressure was 400 bar for 3 s. The temperature of the mold was kept at 50 °C.

10.2.3 Characterization of Nanocomposites

For the characterization of thermally reduced graphene, XRD analysis was conducted on a Panalytical Powder Diffractometer (X'Pert PRO) using Cu-Kα radiation (λ=1.5406 Å, 45 kV, 40 mA). Diffraction pattern was collected in the range of 5-40° 2Θ scale, with a step size of 0.0170° s^{-1}. Raman spectrum was recorded on a Jobin Yvon Horiba LabRAM spectrometer with back-scattered confocal configuration using a HeNe laser (633 nm). A long working distance objective with magnification 50× was used both to collect the scattered light and to focus the laser beam on the sample surface. CCD (charge-coupled device) detector exposure time was 10 s and an average of 2 cycles was used to increase S/N ratio. Raman shifts were calibrated using Si wafer signal at 520.7 cm^{-1}.

Tensile testing of nanocomposites was performed on universal testing machine (Testometric, UK). The sample dimensions for tensile test were: sample length 73 mm, gauge length 30 mm, width 4 mm and thickness 2 mm (conforming to standard ASTM-638-V). A loading rate of 4 mm/min was used and the tests were carried out at room temperature. Tensile modulus and peak stress were calculated using built-in software Win Test Analysis. An average of six values is reported. Dynamic mechanical analysis (DMA) of the polymer materials was performed on RSA3 of TA Instruments to study the change in modulus with temperature (25-100 °C). Dynamic temperature ramp test was used at a frequency of 5 rad/s, ramp rate 5 °C/min and 0.1 % strain for all samples.

Thermal degradation properties of the nanocomposites were recorded using Netzsch thermogravimetric analyzer (TGA). Nitrogen was used as a carrier gas and the scans were obtained from 50 to 700 °C at a heating rate of 10 °C/min. Perkin-Elmer Pyris-1 differential scanning calorimeter under nitrogen atmosphere was used to study the calorimetric properties of nanocomposites. The scans were obtained from 50-200-50 °C (second heating runs) using heating and cooling rates of 5 °C/min respectively. The heat enthalpies were measured with an error of ±0.1 % and were confirmed by repeating the runs.

PRS-801 Resistance system was used to measure the resistance of the samples. It is capable of getting precision resistance measurements from 0.1 up to 2.0e14 Ω with an overall measurement tolerance of ±5 %. Geometric averages of the resistance measured from 3 to 6 different spots from each side of the samples were recorded. Samples were in the shape of circular discs of 2 cm diameter and 1.5 mm thickness. Test voltage was 100 V and test time of 8 sec was used for all samples.

AR 2000 rheometer from TA Instruments was used to measure the rheological performance of the nanocomposites as a function of angular frequency. The measurements were made at 190 °C using a gap opening of 1.6 mm. Strain sweeps were recorded at ω = 1 rad/s from 0.1 to 100% strain and the specimens were observed to be stable up to 10% strain. Frequency sweeps (dynamic testing) were, thus, recorded at 4% strain from ω = 0.1 to 100 rad/s [44].

The morphology of the graphene platelets and nanocomposites was analyzed using transmission electron microscope (EM 912 Omega (Zeiss, Oberkochen BRD)) at 120 kV and 200 kV accelerating voltages. Thin sections of 70-90 nm thickness were microtomed from the block of the nanocomposite specimens. The thin sections or graphene platelets were subsequently supported on 100 mesh grids sputter coated with a 3 nm thick carbon layer.

10.3 Results and Discussion

In the current study, a comparison between the masterbatch (solution mixing followed by melt mixing) and direct melt mixing techniques was made in terms of filler dispersion and properties of the polyethylene-thermally reduced graphene nanocomposites, generated in the presence of compatibilizers. The compounding conditions as well as amounts of composite constituents were held constant during the nanocomposite generation in order to attribute the observed properties solely to compatibilizers and the composite generation technique. The copolymer based compatibilizers used in the study were chosen depending on the functional groups and their amount in the copolymer structure. In order to attain compatibility with the matrix polymer, the non-polar component of the compatibilizers was based on polyethylene. As mentioned in Table 10.1, EAA had 9% acrylic acid content in its structure, whereas the rest was PE. Similarly, EVA-MA was expected to lead to thermodynamic interactions of maleic anhydride group with the filler surface. Ionomer EMAZ was used to study the effect of ionic species in generating interactions with the filler surface. The melt flow indices of the compatibilizers varied in the range of 1-16 g/10 min, thus, indicating a wide variation of their molecular weight characteristics.

The diffraction peak for pure graphene was observed at 26.4° 2Θ, confirming its successful generation. Figure 10.1(a) shows the TEM micrograph of the graphene platelets, whereas Figure 10.1(b) demonstrates the Raman spectrum of these platelets. The Raman spectra of the graphitic materials shows features such as the so-called G band appearing at 1582 cm^{-1} (graphite), the D band at about 1350 cm^{-1}, the D' band at about 1620 cm^{-1} and the G' band at about 2700 cm^{-1}. The evolution of disorder in the generated graphene was quantified using the well-known Tuinstra-Koenig relation $I_D/I_G = C(\lambda)/La$, where I_D/I_G accounts for the intensity ratio between the disorder-induced D band and the Raman-allowed first-order G band. I_D/I_G of the synthesized graphene for this study was observed to be 0.864 (raw peaks without deconvolution) and 0.815 (with deconvolution). In addition, EDX analysis of the graphene surface revealed the C and O fractions to be 77.5% and 18.4%. This indicated the presence of oxygen containing polar surface groups (hydroxyl, epoxide, carboxyl, etc.) in the graphene platelets. Thus, these functional groups present on the filler surface can be expected to interact with the polar compatibilizers used in the study leading to filler delamination in non-polar PE.

Table 10.2 describes the tensile properties of the nanocomposites. Addition of 1% graphene in the absence of any compatibilizers enhanced the tensile modulus by 13%. Addition of 5% EAA to the composite generated by direct melt mixing resulted in 20% increment in modulus, which was further enhanced to 25% for 10% EAA content in the composite. In the case of 5% EAA composite generated by masterbatch approach, much more significant increase of 41%

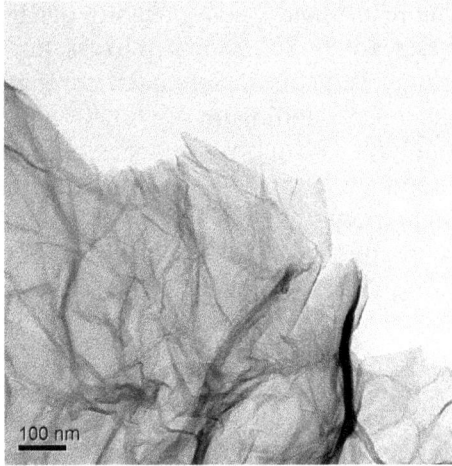

(a)

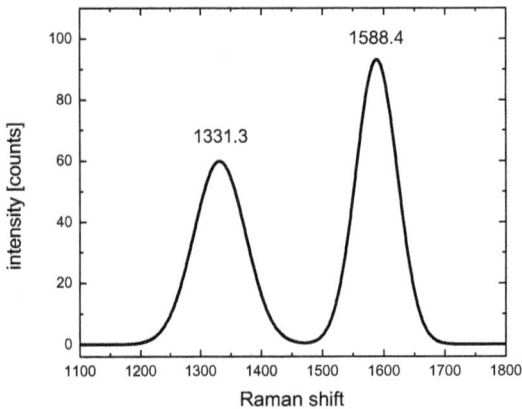

(b)

Figure 10.1 (a) TEM micrograph of the synthesized graphene and (b) Raman spectrum of the graphene sample.

in modulus was observed, indicating this method was able to attain higher degree of filler dispersion. Also, the gain in the modulus did not change with EAA content increasing to 10%, indicating that an efficient mixing of filler with the compatibilizer was obtained at lower compatibilizer content. Peak stress of the MB composites was also marginally better than the corresponding MM composites, though the values were lower than the pure HDPE probably due to

Table 10.2 Tensile properties of nanocomposites (graphene content: 1 wt%, compatibilizer content: 5 and 10 wt%)

Composite	Tensile modulus[a], MPa	Peak stress[b], MPa	Peak strain[c], %	Total elongation[d], mm
HDPE	870	24	7.2	3.4
HDPE/G	982	19	2.7	1.5
HDPE/G/5 EAA MM	1040	22	5.9	3.1
HDPE/G/10 EAA MM	1085	20	3.8	1.1
HDPE/G/5 EMAZ MM	1018	22	6.6	1.3
HDPE/G/10 EMAZ MM	976	20	4.8	1.2
HDPE/G/5 EVA-MA MM	945	22	5.4	1.8
HDPE/G/10 EVA-MA MM	841	20	4.8	1.2
HDPE/G/5 EAA MB	1225	23	4.6	1.3
HDPE/G/10 EAA MB	1222	22	4.4	1.2
HDPE/G/5 EMAZ MB	1207	23	3.5	1.5
HDPE/G/10 EMAZ MB	1011	22	3.1	1.0
HDPE/G/5 EVA-MA MB	1178	22	4.8	1.6
HDPE/G/10 EVA-MA MB	998	21	4.1	1.1

[a]Maximum relative probable error 5%
[b]Maximum relative probable error 15%
[c]Maximum relative probable error 5%
[d]Maximum relative probable error 15%

strain hardening. Reduction in peak strain was also observed in the composites as compared to pure HDPE due to the entrapment of the polymer chains in the filler interlayers and due to presence of some filler aggregates. It also confirmed that the matrix plasticization did not take place on addition of even 10% EAA to the composites. The masterbatch method also led to higher level of increment in the EMAZ composites, however, due to specific nature of the compatibilizer, its 10% content resulted in dominant plasticization effect. EVA-MA composites exhibited the least increment in the modulus. HDPE/G/5 EVA-MA/MM showed an increase of 9% in the modulus as compared to pure polymer. On the other hand, HDPE/G/10 EVA-MA/MM exhibited a slight decrease in the modulus as compared to pure polymer, which signaled poor interactions of the compatibilizer with the filler as well as matrix plasticization. The composites generated by masterbatch approach had 35% and 15% increase in the modulus for 5% and 10% compatibilizer content, which again confirmed the superiority of the method for achieving better filler mixing and property evolution. Thus, though the specific interactions of various compatibilizers with the filler surface resulted in different extents of enhancement in the mechanical performance, the method of nanocomposite generation also played a significant role in enhancing these interactions.

Similar findings were also observed from the DMA analysis shown in Figure 10.2. For instance, at 25 °C, the storage modulus of the composites HDPE/G/5 EAA/MB, HDPE/G/5 EMAZ/MB and HDPE/G/5 EVA-MA/MB respectively was 1.5E9, 1.45E9 and 1.2E9 Pa. The composites also exhibited much higher modulus at higher temperature (>60 °C) as compared to pure polymer as well as HDPE/G composite. This confirmed that the high temperature dimensional stability of the composites was enhanced significantly, though at high temperatures, no significant differences could be observed between the composites with varying compatibilizer contents as well as composites generated by two different methods. Also, some plasticization effects were observed in the composites with 10% compatibilizer content leading to reduction in tensile modulus at room temperatures, however, the higher temperature modulus was still superior than that of pure polymer and HDPE/G composite. This confirmed that the incorporation of the compatibilizers was necessary to enhance the dynamic mechanical response of the nanocomposites, which otherwise would not evolve due to poor dispersion of graphene in HDPE matrix, as observed in HDPE/G composite. Figure 10.3 also represents the loss modulus curves of the composites. In general, the loss modulus of masterbatch composites was higher than the melt mixed composites, due to an increase in internal friction, thus, promoting energy dissipation. The loss peaks indicated the efficiency of the composites to dissipate the mechanical energy. The peaks for all materials exhibited relaxation peak around 40 °C. The magnitude of the peaks for the masterbatch composites

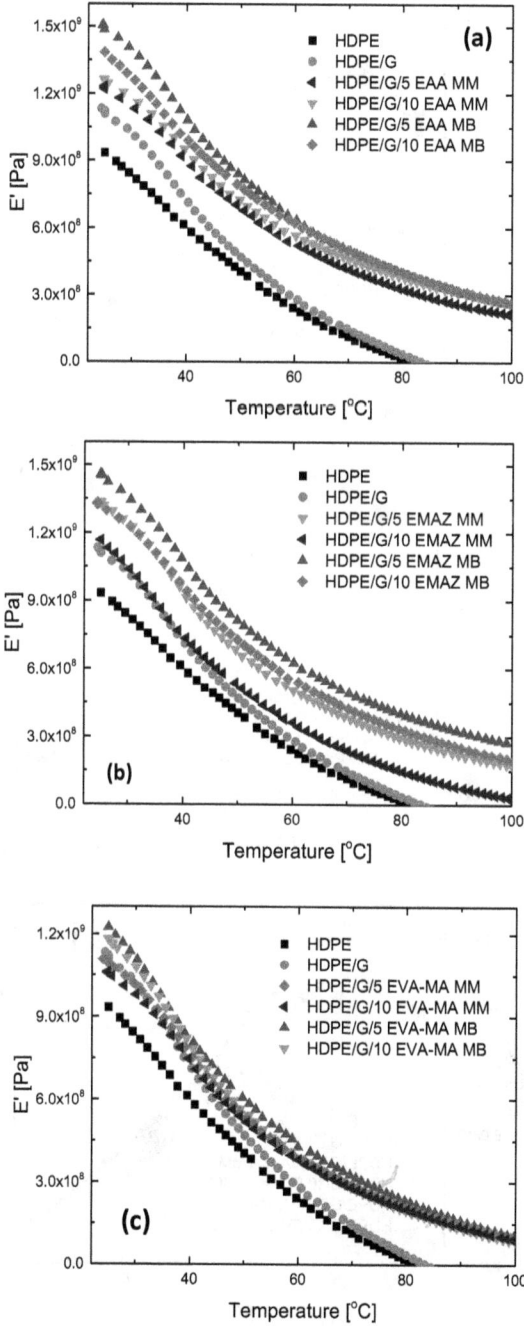

Figure 10.2 Dynamic mechanical profiles of the composites compatibilized with (a) EAA, (b) EMAZ and (c) EVA-MA.

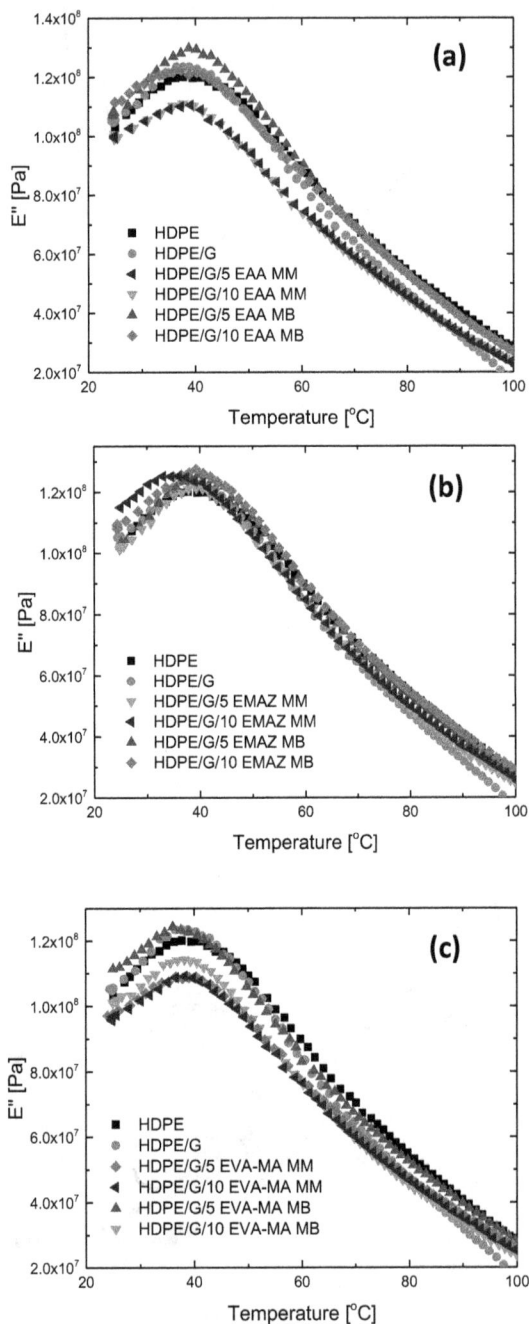

Figure 10.3 Loss modulus profiles of the composites compatibilized with (a) EAA, (b) EMAZ and (c) EVA-MA

was observed to be higher than the melt mixed composites as well as pure polymer. This indicted that the masterbatch composites had higher energy dissipation capacity.

Figure 10.4 also shows the rheological behavior of the nanocomposites as a function of angular frequency. All the composites exhibited higher shear modulus than the pure polymer at all the angular frequencies. The trend in the increment in the shear modulus was also similar to the tensile tests. Composites generated by masterbatch approach and 5% compatibilizer content were observed to have nearly an order of magnitude higher shear modulus than HDPE at lower frequency values. The increment in the modulus was also retained at higher frequencies for EAA and EMAZ compatibilizer systems, but the modulus curves merged with each other for EVA-MA composites at higher frequency values. Complex viscosity of the nanocomposites with 5% compatibilizer fraction is also plotted in Figure 10.5 as a function of angular frequency. Similar to shear modulus, the viscosity of the composites was much higher than the HDPE at all angular frequencies. HDPE/G/5 EAA/MB exhibited the highest increment in viscosity. Moreover, the composites generated with masterbatch approach had much higher viscosity than the composites prepared with direct melt mixing method, also indicating better filler dispersion leading to higher degree of resistance to flow at the same filler content. It is also to be noted that even the viscosity of the composites was enhanced significantly, the processability of these composites was not affected as similar processing conditions as the pure polymer were still suitable for these nanocomposites.

Figure 10.6 compares the morphology of the HDPE/G, HDPE/G/5 EAA/MM and HDPE/G/5 EAA/MB composites at different magnifications. Both composites had much improved filler dispersion as compared to HDPE/G composite [40]. Comparing the two, the composite generated with direct melt mixing exhibited less optimal filler dispersion than the composite generated with masterbatch approach. Filler tactoids of higher thickness were observed for the HDPE/G/5 EAA/MM composite, whereas the tactoid thickness was relatively reduced for the HDPE/G/5 EAA/MB composite. No single exfoliated graphene platelets were observed for the HDPE/G/5 EAA/MM composite, whereas the HDPE/G/5 EAA/MB composite exhibited presence of single graphene layers. Also, the extent of intercalation was observed to be qualitatively higher in the composite generated with masterbatch approach. In addition, no specific filler

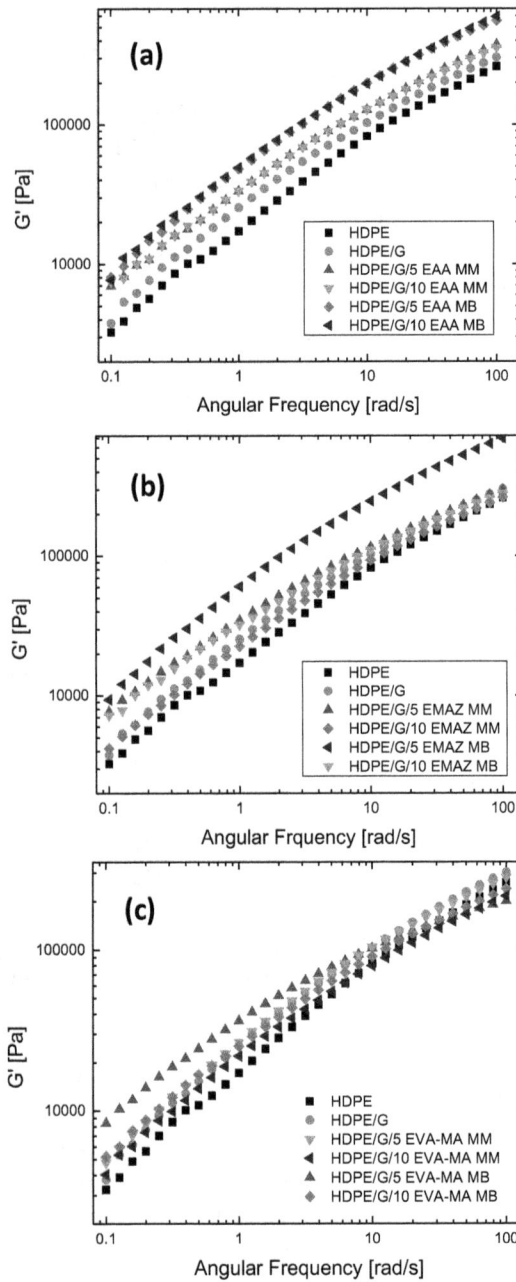

Figure 10.4 Shear modulus of the nanocomposites as a function of angular frequency with (a) EAA, (b) EMAZ and (c) EVA-MA compatibilizers.

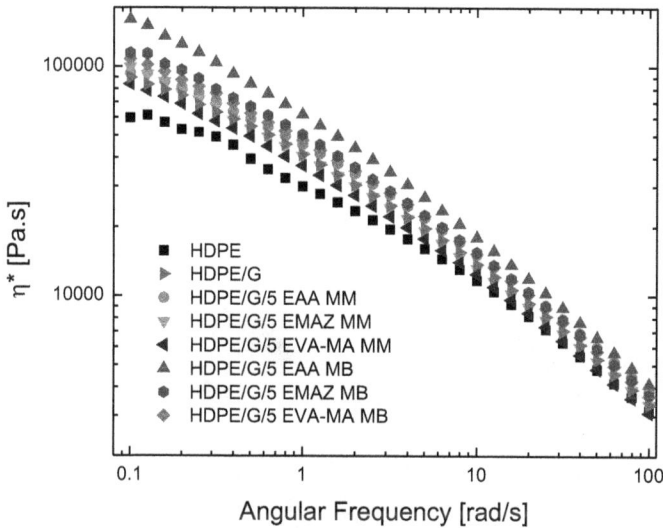

Figure 10.5 Complex viscosity of the nanocomposites as a function of angular frequency.

orientation was observed in the composites at any magnification indicating no flow/stress induced orientation of the filler platelets took place in the molding process. Presence of only 1% filler content is also expected to result in such anisotropic morphology. Other composites also exhibited similar morphology, thus, further confirming the potential of masterbatch method in enhancing the filler dispersion in HDPE matrix. Though no chemical interaction was observed between the compatibilizers and the filler surface, masterbatch method was successful in mixing the filler platelets efficiently with the compatibilizer solution, subsequently enhancing the dispersion of compatibilizer/graphene masterbatch in the HDPE matrix. The findings from tensile, rheological and morphological analysis were further confirmed from the surface resistivity analysis of the samples (Figure 10.7). Pure HDPE exhibited a surface resistivity of 14.01e10 Ωm, which was decreased to 1.04e9 Ωm for HDPE/G composite. HDPE/G/5 EAA/MB, HDPE/G/5 EMAZ/MB and HDPE/G/5 EVA-MA/MB composites exhibited surface resistivity values of 2.08e6, 3.27e7 and 5.38e7 Ωm respectively. On the other hand, HDPE/G/5 EAA/MM, HDPE/G/5 EMAZ/MM and HDPE/G/5 EVA-MA/MM had surface resistivity of 3.75e7, 1.67e8 and 6.39e8 Ωm respectively, indicating better filler dispersion for masterbatch method as well as EAA compatibilizer.

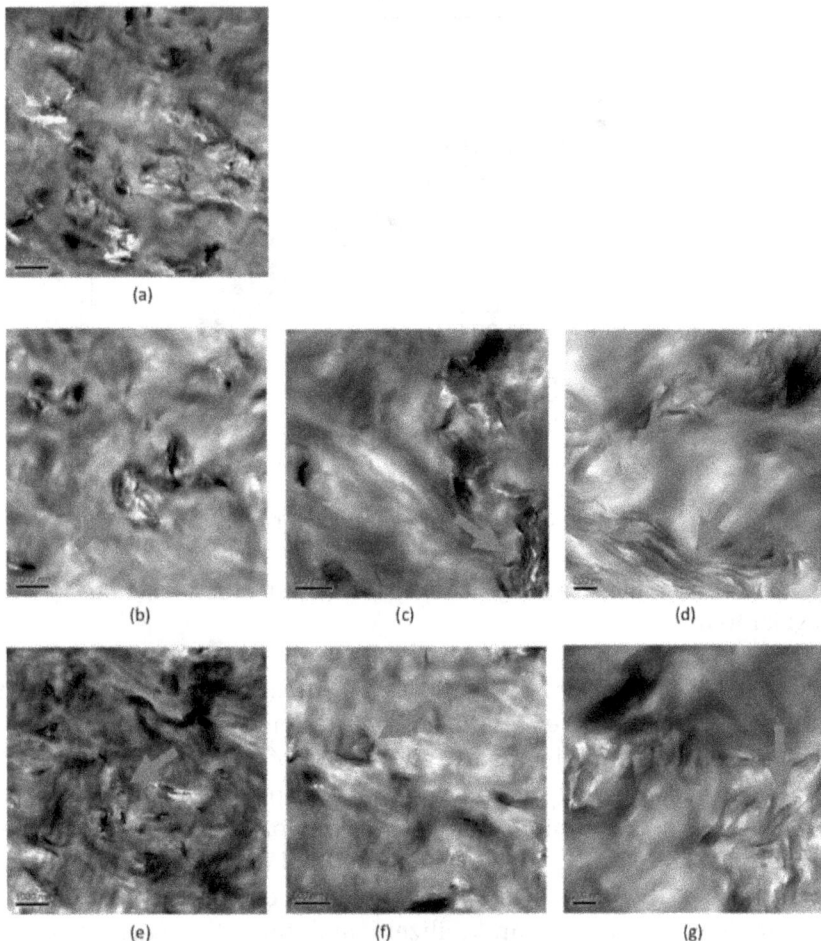

Figure 10.6 TEM micrographs of (a) HDPE/G; (b-d) HDPE/G/5 EAA MM composite at different magnifications and (d-f) HDPE/G/5 EAA MB composite at the same magnifications as (b-d) for comparison.

Table 10.3 describes the calorimetric properties of the nanocomposites. The composites exhibited single broad melting transition, indicating that the compatibilizers were miscible with HDPE. A 3-4 °C decrease in the peak melting temperature was observed in the composites as compared to pure HDPE. In addition, the composites generated with masterbatch approach had 1-2 °C lower peak melting temperatures than the corresponding composites generated by direct melt mixing. Significant changes were observed in the melt enthalpy values (corrected to polymer amount in the composites)

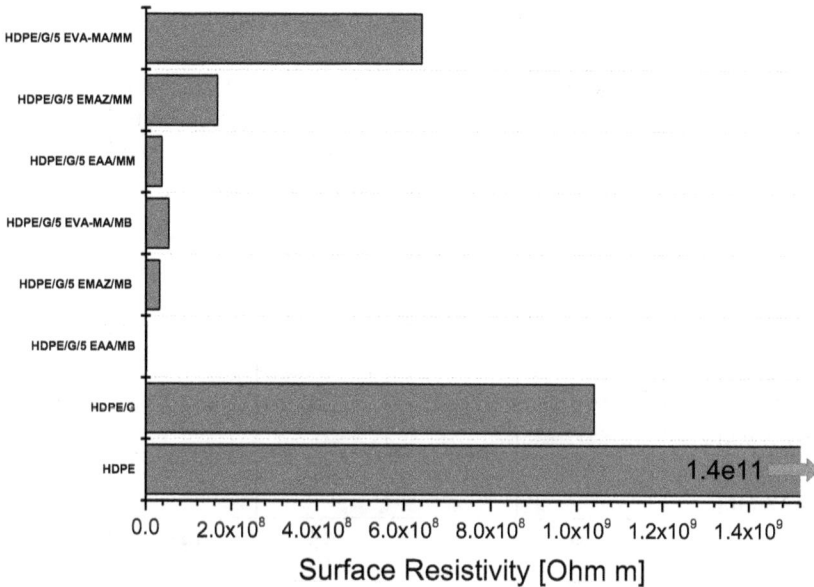

Figure 10.7 Surface resistivity values of the pure polymer and the composites.

depending on the compatibilizer and the composite generation method. HDPE/G/5 EAA/MM and HDPE/G/10 EAA/MM composites exhibited degree of crystallinity of 52% and 41% respectively, as compared to 54% for HDPE. This indicated that the increase in compatibilizer content resulted in hindrance to HDPE crystallization, even though the higher amount of compatibilizer resulted in enhancement in the tensile modulus due to increased extent of filler dispersion. In comparison, HDPE/G/5 EAA/MB and HDPE/G/10 EAA/MB composites had 54% and 45% degree of crystallinity respectively. It again underlined that due to better interaction of the compatibilizer with the graphene platelets in the masterbatch method, overall reduction in the crystallinity at higher compatibilizer fractions was less. Similar observations could also be observed for EMAZ and EVA-MA compatibilizer systems. Peak crystallization temperature of HDPE was also observed to remain constant in the composites, which indicated that the filler platelets did not have any nucleating effect on the polymer. The rheological behavior was also analyzed using criteria reported for compatibility by Han and Chuang [44,45] as shown in Figure 10.8(a,b). When G′ is plotted vs. G″, such analysis generates composition independent correlation for

Table 10.3 Calorimetric properties of the nanocomposites

Composite	Peak melting temperature, °C	Melt enthalpy, J/g	Degree of crystallinity, %	Peak crystallization temperature, °C	Crystallization enthalpy, J/g
HDPE	136	157	54	116	156
HDPE/G	133	149	51	117	151
HDPE/G/5 EAA MM	134	153	52	116	150
HDPE/G/10 EAA MM	134	119	41	116	121
HDPE/G/5 EMAZ MM	135	144	49	117	139
HDPE/G/10 EMAZ MM	134	137	47	116	140
HDPE/G/5 EVA-MA MM	135	128	44	116	128
HDPE/G/10 EVA-MA MM	134	115	39	116	113
HDPE/G/5 EAA MB	133	157	54	117	163
HDPE/G/10 EAA MB	133	132	45	116	138
HDPE/G/5 EMAZ MB	132	155	53	117	159
HDPE/G/10 EMAZ MB	132	139	47	117	137
HDPE/G/5 EVA-MA MB	133	134	46	116	134
HDPE/G/10 EVA-MA MB	133	121	41	116	119

compatible systems, whereas the correlation is composition dependent for incompatible systems. For EAA system, concentration independent correlation was observed except for some deviation at lower frequency values indicating the miscibility of the system. In EVA-MA system, a small degree of deviation was observed for the composites with 10% EVA-MA content. It also indicated that the extent of acrylic acid in EAA was suitable for miscibility with HDPE, however, EVA-MA and EMAZ may have suffered due to the high polar content, which resulted in varying degree of miscibility with HDPE.

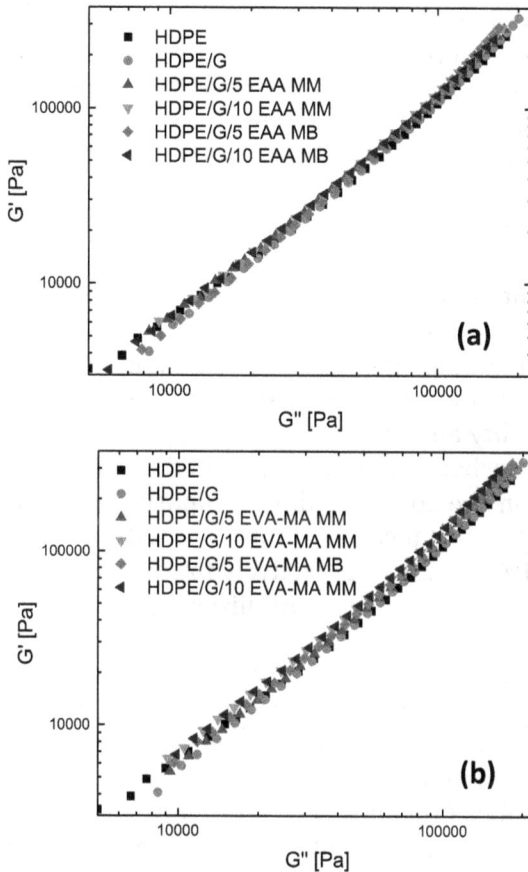

Figure 10.8 Han-Chuang (G' vs. G") plots of (a) EAA and (b) EVA-MA nanocomposites.

In the thermal stability analysis of the nanocomposites, the onset and peak degradation were observed to be significantly delayed in all composites as compared to pure polymer. Due to better filler dispersion, the composites generated with masterbatch method exhibited better thermal performance than the corresponding direct melt mixed composites, thus, resulting in efficient heat transfer to the filler platelets. In addition, increasing compatibilizer content resulted in slight reduction in thermal stability due to low temperature stability of the compatibilizer, however, the degradation still occurred at temperatures higher than the degradation of pure polymer. Synergistic improvement in the thermal stability was observed for HDPE/G/5 EAA/MB and HDPE/G/5 EAA/MM composites as the thermal degradation was even superior to the HDPE/G composite. In other composites, the degradation lied between the HDPE and HDPE/G composites. The dynamic TGA also revealed similar patterns as static TGA, when the time needed to lose 1% of the organic material at fixed temperature was compared.

10.4 Conclusions

In the current study, polyethylene-thermally reduced graphene nanocomposites were generated by direct melt mixing as well as masterbatch (solution mixing followed by melt mixing) methods. The composites were generated using compatibilizers differing in their functionality and amount in the composite. Composites generated with masterbatch approach exhibited much higher mechanical properties than the corresponding composites generated by direct melt mixing. For instance, the composite with 5% EAA had an increment of 41% and 20% respectively for the masterbatch and melt mixed composites. Other compatibilizers showed similar improvements, however, EAA showed the highest extent of property enhancement followed by EMAZ and EVA-MA, thus, confirming the effect of filler-compatibilizer interactions along with the nanocomposite generation method on the resulting properties. Increasing the compatibilizer content to 10% caused matrix plasticization in most of the composites. Composites generated by masterbatch approach and 5% compatibilizer content were observed to have nearly an order of magnitude higher storage modulus than HDPE at lower frequency values. DMA also confirmed that the high temperature dimensional stability of the composites was enhanced significantly on addition of graphene. Morphological studies confirmed the findings from mechanical analysis as extensive delamination was observed in the composites generated by masterbatch approach. Due to better interaction of the compatibilizer with the graphene platelets in the masterbatch method, overall reduction in the crystallinity of HDPE at higher compatibilizer fractions was less. Thermal stability of the nanocomposites also improved on addition of graphene, with 5% EAA composites showing thermal performance even better than the HDPE/G composite.

Summary of the Results

Two routes of nanocomposite generation viz. masterbatch method and direct melt mixing were compared in the current study in relation with the properties of the polyethylene-thermally reduced graphene nanocomposites compatibilized with functional polymers. Filler dispersion was observed to improve in the masterbatch method due to preferential interactions of compatibilizer with the filler, before melt mixing with the matrix polymer. This also translated into improved mechanical properties as more than 2 times increment in tensile modulus was observed in the composite HDPE/G/5 EAA/MB (41%, masterbatch method) as compared to HDPE/G/5 EAA/MM (20%, melt mixing method). In addition, different compatibilizers also influenced the properties to different extent owing to their physical interactions with the filler, however, composites generated with masterbatch were always superior in extent of property enhancements. Increasing the compatibilizer content to 10% also caused varying degree of matrix plasticization. The storage modulus as well as the complex viscosity of the HDPE/G/5 EAA/MB composite was the highest as compared to HDPE/G/5 EMAZ/MB and HDPE/G/5 EVA-MA/MB composites. HDPE/G/5 EAA/MB and HDPE/G/5 EMAZ/MB composites did not show any decrease in degree of crystallinity as compared to pure polymer, whereas the corresponding composites generated with melt mixing exhibited slight decrease in the crystallinity. The composites were observed to have enhanced thermal stability as compared to HDPE. 5% EAA based composites exhibited even higher thermal stability than HDPE/G composites due to improved filler dispersion. Composites generated with masterbatch approach had also higher degree of phase mixing than the corresponding melt mixed composites.

Acknowledgement

The definitive version of this work has been published in Colloid and Polymer Science (2016), 294, 1659-1670. The work has been reproduced here with permission from Springer Nature.

References

1. Alexandre, M., and Dubois, P. (2000) Polymer-layered silicate nanocomposites: Preparation, properties and uses of a new class of materials. *Materials Science and Engineering: R: Reports*, **28**, 1-63.

2. Okada, A., and Usuki, A. (1995) The chemistry of polymer-clay hybrids. *Materials Science and Engineering, C,* **3**(2), 109-115.

3. Osman, M. A., Mittal, V., Morbidelli, M., and Suter, U. W. (2003) Polyurethane adhesive nanocomposites as gas permeation barrier. *Macromolecules,* **36**, 9851-9858.

4. Osman, M. A. Mittal, V. Morbidelli, M. and Suter, U. W. (2004) Epoxy-layered silicate nanocomposites and their gas permeation properties. *Macromolecules.* **37**, 7250-7257.

5. Brechet, Y., Cavaille, J. Y., Chabert, E., Chazeau, L., Dendievel, R., Flandin, L., and Gauthier, C. (2001) Polymer based nanocomposites: Effect of filler-filler and filler-matrix interactions. *Advanced Engineering Materials,* **3**, 571-577.

6. Pavlidoua, S., and Papaspyrides, C. D. (2008) A review on polymer–layered silicate nanocomposites. *Progress in Polymer Science,* **33**, 1119-1198.

7. Mark, J. E. (1996) Ceramic-reinforced polymers and polymer-modified ceramics. *Polymer Engineering & Science,* **36**, 2905-2920.

8. Liu, H., and Brinson, L. C. (2008) Reinforcing efficiency of nanoparticles: A simple comparison for polymer nanocomposites. *Composites Science and Technology,* **68**, 1502-1512.

9. *Advances in Diverse Industrial Applications of Nanocomposites,* Reddy, B. (ed.), InTech, UK (2011).

10. Hui, L., Smith, R. C., Wang, X., Nelson, J. K., and Schadler, L. S. (2008) Quantification of Particulate Mixing in Nanocomposites, *2008 Annual Report Conference on Electrical Insulation and Dielectric Phenomena,* Canada, doi: 10.1109/CEIDP.2008.4772831.

11. Manas-Zloczower, I., and Cheng, H. (1996) Analysis of mixing efficiency in polymer processing equipment. *Macromolecular Symposia,* **112**(1), 77-84.

12. Lee, S. H., Cho, E., Jeon, S. H., and Youn, J. R. (2007) Rheological and electrical properties of polypropylene composites containing functionalized multi-walled carbon nanotubes and compatibilizers. *Carbon,* **45**, 2810-2822.

13. Serageldin, M. A., and Wang, H. (1987) Effect of operating parameters on time to decomposition of high density polyethylene and chlorinated polyethylenes. *Thermochimica Acta,* **117**, 157-166.

14. Lee, C., Wei, X., Kysar, J. W., and Hone, J. (2008) Measurement of the elastic properties and intrinsic strength of monolayer graphene. *Science,* **321**, 385-388.

15. Zhao, H., Min, K., and Aluru, N. R. (2009) Size and chirality dependent elastic properties of graphene nanoribbons under uniaxial tension. *Nano Letters,* **9**, 3012-3015.

16. Lier, G. V., Alsenoy, C. V., Doren, V. V., and Greelings, P. (2000) Ab initio study of the elastic properties of single-walled carbon nanotubes and graphene. *Chemical Physics Letters,* **326**, 181-185.

17. Robertson, D. H., Brenner, D. W., and Mintmire, J. W. (1992) Energetics of nanoscale graphitic tubules. *Physical Review B*, **45**, 12592.

18. Itkis, M. E. Borondics, F. Yu, A. Haddon, R. C. (2007) Thermal conductivity measurements of semitransparent single-walled carbon nanotube films by a bolometric technique. *Nano Letters*, **7**, 900-904.

19. Park, S., and Rouff, S. (2009) Chemical methods for the production of graphenes. *Nature Nanotechnology*, **4**, 217-224.

20. Song, M., and Cai, D. (2012) Graphene functionalization: A review. In Polymer Graphene Nanocomposites, Mittal, V. (ed.)., Royal Society of Chemistry, UK, pp. 1-51.

21. Mukhopadhyay, P., and Gupta, R. K. (2011) Trends and frontiers in graphene- based polymer nanocomposites. *Plastics Engineering*, **67**, 32-42.

22. Geim, A. K., and Novoselov, K. S. (2007) The rise of graphene. *Nature Materials*, **6**, 183-191.

23. Compton, O. C., and Nguyen, S. B. T. (2010) Graphene oxide, highly reduced graphene oxide, and graphene: versatile building blocks for carbon-based materials. *Small*, **6**, 711-723.

24. Mittal, V. (2014) Functional polymer nanocomposites with graphene: A review. *Macromolecular Materials and Engineering*, **299**, 906-931.

25. Mittal, V., Luckachan, G. E., and Matsko, N. B. (2014) PE/chlorinated-PE blends and PE/chlorinated-PE/graphene oxide nanocomposites: Morphology, phase miscibility, and interfacial interactions. *Macromolecular Chemistry and Physics*, **215**, 255-268.

26. Mittal, V., Chaudhry, A. U., and Luckachan, G. E. (2014) Biopolymer–thermally reduced graphene nanocomposites: structural characterization and properties. *Materials Chemistry and Physics*, **147**, 319-332.

27. Supova, M., Martynkova, G. S., and Barabaszova, K. (2011) Effect of nanofillers dispersion in polymer matrices: A review. *Science of Advanced Materials*, **3**, 1-25.

28. Kuilla, T., Bhadra, S., Yao, D., Kim, N. H., Bose, S., and Lee, J. H. (2010) Recent advances in graphene based polymer composites. *Progress in Polymer Science*, **35**, 1350-1375.

29. Fang, M., Wang, K., Lu, H., Yang, Y., and Nutt, S. (2009) Covalent polymer functionalization of graphene nanosheets and mechanical properties of composites. *Journal of Materials Chemistry*, **19**, 7098-7105.

30. Stankovich, S., Piner, R. D., Chen, X., Wu, N., Nguyen, S. T., and Ruoff, R. S. (2006) Stable aqueous dispersions of graphitic nanoplatelets via the reduction of exfoliated graphite oxide in the presence of poly (sodium 4-styrenesulfonate). *Journal of Materials Chemistry*, **16**, 155-158.

31. Stankovich, S., Dikin, D. A., Dommett, G. H. B., Kohlhaas, K. M., and Zimney, E. J. (2006) Graphene-based composite materials. *Nature*, **442**, 282-286.

32. Castelain, M., Martinez, G., Marco, C., Ellis, G., and Salavagione, H. J. (2013) Effect of click-chemistry approaches for graphene modification on the electrical, thermal, and mechanical properties of polyethylene/graphene nanocomposites. *Macromolecules*, **46**, 8980-8987.

33. Vasileiou, A. A., Kontopoulou, M., and Docoslis, A. (2014) A noncovalent compatibilization approach to improve the filler dispersion and properties of polyethylene/graphene composites. *ACS Applied Materials & Interfaces*, **6**, 1916-1925.

34. Yu, B., Wang, X., Qian, X., Xing, W., Yang, H., Ma, L., Lin, Y., Jiang, S., Song, L., Hu, Y., and Lo, S. (2014) Functionalized graphene oxide/phosphoramide oligomer hybrids flame retardant prepared via in situ polymerization for improving the fire safety of polypropylene. *RSC Advances*, **4**, 31782-31794.

35. Seo, H. M., Park, J. H., Dao, T. D., and Jeong, H. M. (2013) Compatibility of functionalized graphene with polyethylene and its copolymers. *Journal of Nanomaterials*, **2013**, Article ID 805201.

36. Wang, J., Xu, C., Hu, H., Wan, L., Chen, R., Zheng, H., Liu, F., Zhang, M., Shang, X., and Wang, X. (2011) Synthesis, mechanical, and barrier properties of LDPE/graphene nanocomposites using vinyl triethoxysilane as a coupling agent. *Journal of Nanoparticle Research*, **13**, 869-878.

37. Kim, S., Do, I., and Drzal, L. T. (2009) Multifunctional xGnP/LLDPE nanocomposites prepared by solution compounding using various screw rotating systems. *Macromolecular Materials and Engineering*, **294**, 196-205.

38. Ansari, S., and Giannelis, E. P. (2009) Functionalized graphene sheet-Poly (vinylidene fluoride) conductive nanocomposites. *Journal of Polymer Science, Part B: Polymer Physics*, **47**, 888-897.

39. Potts, J. R., Dreyer, D. R., Bielawski, C. W., and Ruoff, R. S. (2011) Graphene-based polymer nanocomposites. *Polymer*, **52**, 5-25.

40. Chaudhry, A. U., and Mittal, V. (2013) High-density polyethylene nanocomposites using master batches of chlorinated polyethylene/graphene oxide. *Polymer Engineering and Science*, **53**, 78-88.

41. McAllister, M. J., Li, J. L., Adamson, D. H., Schniepp, H. C., Abdala, A. A., Liu, J., Herrera-Alonso, M., Milius, D. L., Car, R., Prudhomme, R. K., and Aksay, I. A. (2007) Single sheet functionalized graphene by oxidation and thermal expansion of graphite. *Chemistry of Materials*, **19**, 4396-4404.

42. Hummers, W. S., and Offeman, R. E. (1958) Preparation of graphitic oxide. *Journal of the American Chemical Society*, **80**, 1339.

43. Mittal, V., and Chaudhry, A. U. (2015) Polymer-graphene nanocom-

posites: Effect of polymer matrix and filler amount on properties. *Macromolecular Materials and Engineering*, **300**(5), 510-521.

44. Chuang, H. K., and Han, C. D. (1984) Rheological behavior of polymer blends. *Journal of Applied Polymer Science*, **29**, 2205-2229.

45. Han, C. D., and Chuang, H. K. (1985) Criteria for rheological compatibility of polymer blends. *Journal of Applied Polymer Science*, **30**, 4431-4454.

Index

■ Index

P

packaging, 79, 160, 192, 238
paraffin wax, 147, 158
peak crystallization
  temperature, 220, 259
peak degradation, 221, 261
peak intensity, 191
peak melting temperature,
  50, 258
penetration, 106, 181
percolation, 111-112, 186,
  188, 243
permeability, 42, 70, 155, 160
PES, 3, 135
phase miscibility, 42, 68,
  193, 265
physical mixing, 28
plasticization, 12, 37-38, 185,
  224, 251-252, 262-263
polar groups, 1
polar polymers, 48, 179
polarization, 75-76, 115
polyacrylamide, 43
polyaniline, 28, 200, 210
polycondensation, 77, 79
polyesters, 145, 147, 170, 197
polyimide, 28, 43, 77-78,
  93-95, 198, 239
polymer-filler interface, 179

polymerization, 72, 77-78,
  81-82, 84-89, 95-96, 152,
  197, 199, 203-204, 209, 239,
  242, 266
polyolefin, 3, 21-22, 28, 71,
  136, 144, 148, 170, 172, 175,
  241, 243
polystyrene, 2, 24, 42, 87-88,
  98, 128, 162, 170, 176, 199,
  216, 238, 242
polyurethane, 3, 22, 27, 29,
  43-44, 111, 120, 143, 148,
  153, 165, 170, 173, 175, 198
precursor, 77, 79-80, 84, 147,
  202, 212-213, 244
processability, 47-48
processing temperature,
  148, 155, 160, 219, 231, 247
PVB, 42
PZT, 114

R

Raman spectra, 8-9, 51,
  62-63, 249
recyclability, 124
reinforcement, 25, 44, 73,
  106, 111, 129, 141, 241-242
relaxation, 86, 252
rheological properties, 21,

W

U

X

V

Y

www.ingramcontent.com/pod-product-compliance
Lightning Source LLC
Chambersburg PA
CBHW050455190326
41458CB00005B/1294